CAMBRIDGE LIBRARY COLLECTION

Books of enduring scholarly value

Darwin

Two hundred years after his birth and 150 years after the publication of 'On the Origin of Species', Charles Darwin and his theories are still the focus of worldwide attention. This series offers not only works by Darwin, but also the writings of his mentors in Cambridge and elsewhere, and a survey of the impassioned scientific, philosophical and theological debates sparked by his 'dangerous idea'.

A Dictionary of Botanical Terms

John Stevens Henslow (1796 – 1861) was a botanist and geologist. As teacher, mentor and friend to Charles Darwin, it was his introduction that secured for Darwin the post of naturalist on the voyage of the Beagle. While Professor of Botany, Henslow established the Cambridge University Botanic Garden as a resource for teaching and research. Students were encouraged to examine plant specimens carefully, and to record the characteristics of their structures. Henslow would have known how daunting they found the task of becoming proficient with botanical vocabulary, and produced this volume to provide a secure foundation for scientific investigations. This meticulous glossary, originally published as a single volume in 1857 but drawing on contributions he made earlier to issues of The Botanist and Maund's Botanic Garden, is a testament to Henslow's scholarship. It is liberally illustrated with delightful woodcuts that clarify the meaning of selected terms.

Cambridge University Press has long been a pioneer in the reissuing of out-of-print titles from its own backlist, producing digital reprints of books that are still sought after by scholars and students but could not be reprinted economically using traditional technology. The Cambridge Library Collection extends this activity to a wider range of books which are still of importance to researchers and professionals, either for the source material they contain, or as landmarks in the history of their academic discipline.

Drawing from the world-renowned collections in the Cambridge University Library, and guided by the advice of experts in each subject area, Cambridge University Press is using state-of-the-art scanning machines in its own Printing House to capture the content of each book selected for inclusion. The files are processed to give a consistently clear, crisp image, and the books finished to the high quality standard for which the Press is recognised around the world. The latest print-on-demand technology ensures that the books will remain available indefinitely, and that orders for single or multiple copies can quickly be supplied.

The Cambridge Library Collection will bring back to life books of enduring scholarly value (including out-of-copyright works originally issued by other publishers) across a wide range of disciplines in the humanities and social sciences and in science and technology.

A Dictionary of Botanical Terms

John Stevens Henslow

CAMBRIDGE UNIVERSITY PRESS

Cambridge, New York, Melbourne, Madrid, Cape Town, Singapore,
São Paolo, Delhi, Dubai, Tokyo

Published in the United States of America by Cambridge University Press, New York

www.cambridge.org
Information on this title: www.cambridge.org/9781108001311

This edition first published 1856
This digitally printed version 2009

ISBN 978-1-108-00131-1 Paperback

A
DICTIONARY
OF
BOTANICAL TERMS.

BY

THE REV. J. S. HENSLOW M.A.,

PROFESSOR OF BOTANY IN THE UNIVERSITY OF CAMBRIDGE.

ILLUSTRATED

BY NEARLY TWO HUNDRED CUTS.

LONDON:

GROOMBRIDGE AND SONS,

PATERNOSTER-ROW.

PREFACE.

THIS Dictionary was published at intervals; portions being appended successively to the monthly numbers of Maund's Botanist, and Maund's Botanic Garden. It contains a copious list of the Latin and English terms which have been used by various Botanical Authors, the former distinguished by Italic Capitals, the latter by Roman Capitals. The Greek or Latin derivatives are in brackets, immediately after the terms. To avoid needless repetitions, (when the case admits) reference is made from the Latin to the corresponding English term, where the explanation is alone given; small wood cuts occasionally assist in illustrating some of the terms. The names of the Natural Orders are also given, and these are referred to their Classes. It was originally intended to give short accounts of the Orders, (as under Acanthaceæ, Acerineæ, &c.,) but as these seemed needlessly to increase the quantity of matter, and were not exactly in harmony with the technical character of the Dictionary, they were early discontinued. A science with a technical terminology of about 2000 words and synonymes may appear repulsive; but a little consideration will satisfy us that this need not be the case. A large number of the terms here recorded have been very needlessly employed, and are only met with in the works of the older botanists.

Many of the words employed in describing plants retain their ordinary acceptation, and others which have a more technical application need to be explained only once to be easily retained. Those words which have been exclusively coined for this science, and are still in use, are not so numerous as to alarm the least energetic of its votaries. Such a Dictionary as we now present in its complete state, affords every one a ready reference to any term that may be met with in botanical authors, whether it be still in use, or has become so far obsolete that even proficients in the science may be at a loss to ascertain its meaning without more trouble than they would be willing to bestow. No slight confusion occurs in the minds of beginners from the different sense in which different authors have sometimes employed the same word; and also from the identity in meaning which they have attached to different words. Carrying on their labours independently, and finding it necessary to give expression to some newly observed fact, authors have done this in ignorance that another observer may be doing the same thing at the same time, or may have done it before. What often happens with respect to names given nearly simultaneously to the same plant by different describers, or given by one in ignorance of the labours of another, has occurred to an unfortunate extent in botanical Terminology; and hence we are overloaded with synonymes. It is here especially that our Dictionary will be found serviceable; let any one turn to the word "Receptacle," and he will appreciate this remark. By observing in which of its significations a particular Author employs a special term, all doubt as to his meaning is immediately at an end.

It is certainly to the difficulties which the undue extension of our botanical nomenclature has thrown in the way of beginners, that so many are inclined to turn aside from systematic botany, and to direct their attention, too exclusively, to the engaging speculations of botanical physiology. Without doubt physiology is the higher department of the science, and minute vegetable anatomy a branch of investigation essential to its progress. But it is in vain to attempt raising a superstructure that will be likely to stand, until the foundations shall have been securely laid. And assuredly the labours of systematic botanists, in the present state of our science, are those most needed, and will be so for some time to come, or there will be no steady progress for Botany. The truly scientific systematist is far from avoiding the investigations of the vegetable anatomist and physiologist. No sure step in advance is now to be made in systematic botany without careful dissections, and some reference to the functions of specific organs. All must remain vague and unsatisfactory in physiology which is not secured by those bonds, (constantly strengthening) by which System combines all clearly-ascertained "Facts," and gives expression to the nearest approximation we can hope to make to the Divine scheme upon which this portion of the Creation has been constructed.

J. S. HENSLOW.

A

DICTIONARY,

OF

ENGLISH AND LATIN TERMS,

USED IN

BOTANICAL DESCRIPTIONS.

The explanations will be given under the several English terms, and the Latin terms will be printed in Italics, generally, with a mere reference to the corresponding English ones.

A (from the Greek *a*) in composition, signifies privation, or absence of the object expressed. Thus, *APHYLLUS*, leafless; *ACAULIS*, stemless. If the word to which it is prefixed begin with a vowel, it is softened into *AN;* thus, *ANANTHUS*, flowerless.

ABBRE'VIATED, (*AB* from, *BREVIS* short) when an organ, or part of an organ, is shorter than another to which it is contiguous.

ABBREVIA'TUS, abbreviated.

ABER'RANT, (*AB* from, *ERRO* to wander) where the characters of certain species or groups differ materially from those of others, to which they are most nearly related.

ABIE'TINUS, (*ABIES* spruce-fir) used for designating certain cryptogamic plants which grow on evergreen trees.

ABNOR'MAL, (*AB* from, *NORMA* law) deviating from regularity, natural condition, or more usual structure of other allied species.

ABNORMA'LIS, abnormal.

ABORI'GINAL, (*AB* from, *ORIGO* a beginning) plants which appear to be the spontaneous production of any country. The same as indigenous.

ABOR'TIENS, abortive.

No. 1.

ABOR'TION, (*AB* from, *ORIOR* to rise, to be born) the suppression or absence of an organ, arising from its non-development. Its actual existence is either assumed by analogy, or is sometimes detected by an accidental or monstrous condition of a plant. Thus, in those varieties of the two genera, *ANTIRRHINUM* and *LINARIA*, which are termed *PELORIA*, (i. e. monstrous) a fifth stamen is developed, and the corolla becomes regular, fig. 1, instead of being personate and didynamous, fig. 2.

1

2

ABOR'TIVE, defective, barren. See abortion.

ABORTI'VUS, abortive.

ABOR'TUS, abortion.

ABRUPT, (*AB* from, *RUMPO* to break) when some part appears as if it were suddenly terminated.

ABRUPT'LY-PINNATE, Where a pinnate leaf is without an odd leaflet at its extremity, as fig. 3.

3

ABRUP'TUS, abrupt.

AB'SOLUTE, (*AB* from, *SOLVO* to loose) applied to the insertion of an organ, with respect to its actual position; in contradistinction to its relative position with other organs. Thus, when the stamens in a rose are said to be perigynous, this term marks their position relatively, with respect to the pistils; but when the rose is said to be calycifloral, the absolute position of the stamens is alluded to, as being placed on the calyx.

ABSOLU'TUS, absolute.

ABSORP'TION, (*ABSORBEO* to suck in) the function by which the spongioles imbibe the moisture which becomes sap.

ACALYCA'LIS, (*a* without, *καλυξ* a calyx) where the stamens contract no adhesion with the calyx.

ACALYCI'NUS, *ACAL'YCIS*, (*a* without, *καλυξ* a calyx) where the calyx is wanting.

ACANTHA'CEÆ, OR ACAN'THI, *JUSSIEU*. (from the genus ACANTHUS) the Justicia tribe. A natural order, of which the most usual and prominent characteristics are, an irregular two-lipped corolla, much resembling that of some Labiatæ; with the stamens didynamous, but generally reduced to two, by the total or partial abortion of one pair. The ovary is two-celled, and the capsule opens elastically with a loculicidal dehiscence. No albumen. The species are chiefly tropical herbs and shrubs, with opposite leaves.

CANTHOCAR'PUS (ἄκανθα a thorn, καρπὸς fruit) where a fruit is furnished with spines.

ACANTHOCLA'DUS (ἄκανθα a thorn, κλάδος a branch) where the branches are furnished with spines.

ACANTH'OPHORUS (ἄκανθα a thorn, φέρω to bear) furnished with spines, or large stiff bristles.

ACANTHOPO'DIUS, (ἄκανθα a thorn, ποῦς a foot) where the petioles or footstalks to the leaf are furnished with spines.

ACAU'LIS, (*a* without, *CAULIS* a stem) stemless.

ACCES'SORY, (*ACCESSUS* an increase) something superadded to the usual condition of an organ

ACCIS'US, (cut or clipt) where the extremity appears as if it were cut away; much the same as truncate.

ACCLI'MATIZE, (*AD* to, *CLIMA* a climate) to accustom a plant to live in the open air without protection, in a country where it is not indigenous.

ACCRES'CENS, (*AD* to, *CRESCO* to grow) persistent and increasing in size, as the calyx of Physalis alkakengi; the styles of Anemone pulsatilla, &c.

ACCRETE, (*AD* to, *CRESCO* to grow) when contiguous parts or organs become naturally grafted together.

ACCRE'TUS, accrete.

ACCUM'BENT, (*AD* to, *CUBO* to lie down) when one part lies close upon the edge of another; as where the radicle is bent round and pressed against the edges of the cotyledones, in certain Cruciferæ fig. 4. The symbol (○═) is frequently made use of to signify this term. It is used in opposition to "incumbent."

ACEPH'ALOUS, (*a* without, κεφαλη a head) when the style does not stand on the summit of the ovary, but proceeds from the side, or near the base, fig. 5.

ACEPH'ALUS, acephalous.

ACERELLA'TUS, somewhat acerose.

ACERINEÆ, (from the genus ACER) the Sycamore tribe. A small natural Order composed of trees peculiar to the more temperate parts of the northern hemisphere. The flowers are usually small and green, and generally contain both calyx and corolla, varying in the number of their parts from four to nine. The stamens spring from an hypogynous disk and are about eight in number. The flowers are occasionally polygamous. The ovary is two-lobed, and the fruit possesses the peculiar winged structure termed a Samara.

A'CEROSE, (*ACUS* a needle) linear and sharp-pointed. Applied especially to the leaves of the Fir-tribe; fig. 6.

A'CEROSUS, acerose.

ACETAB'ULOUS, ACETAB'ULIFORM, (*ACETABULUM* a cup, *FORMA* shape) shaped like a cup or saucer; as the fructification on many lichens, fig. 7.

ACETABULIFOR'MIS, acetabuliform.

ACETABULO'SUS, acetabulous.

ACETA'RIUS, (*ACETARIA* salad) suited for salads.

ACHASCOPH'YTUM, (α not, χάσκω to open, φυτον a plant) a plant which has an indehiscent fruit.

ACHE'NIUM, (α not, χάινω to open) this term is applied, by different authors, to two distinct kinds of fruit. 1. Where the fruit is superior, and consequently the pericarp is not invested by the calyx. It is dry, hard, single-seeded, and indehiscent. This is otherwise termed a Nut. 2. Where the pericarp is inferior, and consequently invested by the calyx; in other respects resembling the last, but usually not so hard. The seeds of compositæ are the best examples, fig. 8.

ACHENO'DIUM, a fruit composed of two or more achenia, as in the umbelliferæ. More usually called "cremocarpium."

ACHLAMYD'EOUS, (α without, χλαμις a coat) flowers without any distinct perianth; as in the willows, where the stamens or pistil are merely subtended by a bractea, fig. 9.

ACHYROPH'YTUM, αχυρον chaff, φυτον a plant) a plant having glumaceous flowers.

ACIDIF'EROUS, (*ACIDUM* an acid, *FERO* to bear) containing some acid principle.

ACIDO'TUS, (ἀκιδωτὸς pointed) when the branches or other organs terminate in a spine, or hard point.

A'CIES, an edge formed by the intersection of two planes. More often termed an "angle," in stems, fruit, &c.

ACIC'ULA, (diminutive of *ACUS* a needle) a name given to the rachis of some grasses, where it is reduced to a mere bristle.

ACIC'ULAR, (*ACUS* a needle) of a slender form, like a needle.

ACICULA'RIS, acicular.

ACICULA'TED, (*ACUS* a needle) superficially marked, as if irregularly scratched with the point of a needle.

ACICULA'TUS, *ACICULI'NUS*, aciculated.

ACINACIFO'LIUS, (*ACINACES* a scymiter, *FOLIUM* a leaf) a fleshy leaf, curved like a scymiter, with a thin edge and broad back, fig. 10.

ACINA'CIFORM, (*ACINACES* a scymiter, *FORMA* shape) formed like a scymiter.

ACINACIFOR'MIS, acinaciform.

ACINA'RIUS, (*ACINUS* the seed of grapes) when a stem or branch is covered with little spherical and stalked vesicles, looking like grape seeds; as in some sea-weeds.

ACINODEN'DRUS, (ἄκινος grape-seed, δενδρον a tree) a plant whose fruit is arranged in bunches.

ACINO'SUS, (*ACINUS* grape-seed) shaped like the seed of a grape.

ACI'NUS, (ἄκινος grape-seed) not applied in its classical sense to the actual seed; but employed to signify the berries which compose the bunch of grapes, or other pulpy berries containing hard seeds, as the single granules of which the raspberry is composed.

ACIPH'YLLUS, (ἀκὴ a point, φύλλον a leaf) a linear and pointed leaf, fig. 11.

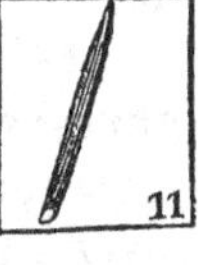

ACLYTHROPH'YTUM, (α without, κλεῖθρον a door, φυτὸν a plant) plants whose seeds are supposed to be naked, or without a pericarp.

A'CORN, see *GLANS*.

ACOROI'DEÆ, ACORA'CEÆ OR ACORI'NÆ, (from the genus ACORUS) a natural group which may either be considered as a distinct order, or as a tribe of the order AROIDEÆ, from the rest of which it differs more particularly in habit and in the presence of the scaly rudiment of the perianth.

ACOTYLE'DONOUS, wanting cotyledons. See Acotyledons.

ACOTYLE'DON, (α without κοτυληδων a seed leaf) a plant belonging to those flowerless tribes, which have no true seeds, but are reproduced by sporules. Otherwise, termed a cryptogamic plant.

ACOTYLE'DONES, used as a synonyme for *CRYPTOGAMIA* by some botanists; whilst others consider that a portion of the latter, as the ferns, are really monocotyledonous. As synonymous with the Linnean class Cryptogamia, the Acotyledones form a natural class, which includes all the flowerless plants; and is sub-divided into several very distinct orders, as 1, Filices, (Ferns) 2, Lycopodiaceæ (Club-mosses) 3,

Equisetaceæ: 4, Musci, (Mosses) 5, Hepaticæ, (Liverworts) 6, Characeæ: 7, Algæ, (Sea-weeds) 8, Lichenes: 9, Fungi, (Mushrooms).

AC'ROGEN, (ἄκρος the extremity, γεννάω to produce) a name given to cellular, or cryptogamic plants, in reference to the manner in which their stems increase, by additions to the extremity merely: and not by the formation of new matter, internally, or externally, throughout their whole length, as in endogens and exogens.

ACRO'NYCHIUS, (ἄκρος a summit, ονυξ a claw) curved like the claw of an animal.

ACROSAR'CUM, (ἄκρος a summit, σὰρξ flesh) a spherical fleshy fruit, adhering to the calyx, by whose limb it is often crowned; as in currants. Synonyme for Berry.

ACROSPI'RA, (ἄκρος a summit, σπεῖρα a chord) a name which has been given to a plumule, as in the barley, which in germination rises like a chord from the summit of the seed.

ACTINEN'CHYMA, (ἀκτὶν a ray of light, χύμα juice) the cellular tissue which forms the medullary rays.

ACTINOCAR'PUS, (ἀκτὶν a ray, καρπος fruit) where the pla centæ are ranged in a radiated manner.

ACTINOSTO'MUS, (ἀκτὶν a ray, ςομα mouth) the radiated structure sometimes observable round the little openings termed *OSTIOLA*, on the frond of Algæ, the thallus of Lichenes, &c.

ACU'LEATE, (*ACULEUS* a prickle) sharply pointed; also, prickly.

ACULÆA'TUS, *ACULEIFORMIS*, aculeate.

ACULEO'SUS, furnished with prickles.

ACU'LEUS, a prickle.

ACU'MEN, a tapering point.

ACU'MINATE, ACU'MINATED, (*ACUMEN* a point) ending in a long taper point.

ACUMINATELY-CUSPIDATE. Acuminate, and ending in a bristle.

ACUMINA'TUS acuminate.

ACUMINIFO'LIUS, (*ACUMEN* a point, *FOLIUM* a leaf) where the leaf is acuminate.

ACU'MINOSE, approaching to acuminate.

ACUTAN'GULAR, (*ACUTUS* sharp, *ANGULUS* an angle) where the edges of stems, &c. are sharp, and a transverse section presents acute angles; fig. 12. Sometimes used also, where the leaves are divided into many narrow lobes.

12

Acutan'gulus, acutangular.

Acu'te, (*acutus* sharp) where the extremities present an angle less than a right angle.

Acu'te-emargina'tus, notched, but ending abruptly.

Acutiflo'rus, (*acutus* sharp, *flora* a flower) where the petals, or lobes of the corolla, terminate in a point.

Acutifo'lius, (*acutus* sharp, *folium* a leaf) where the leaves are pointed.

Acutilo'bus, (*acutus* sharp, *lobus* a lobe) where the lobes of the leaves are pointed.

Acutius'culus, somewhat acute.

Acu'tus, acute.

Addi'tional-mem'brane, same as embryonic sack.

Adducto'res, (*ad* to, *duco* to lead) the young state of the *thecæ* of mosses. These being crowded together are mostly abortive, whilst one only is usually developed, at least at the same spot.

Adel'phic, Adelphous, (αδελφος a brother) when the stamens are united by their filaments into one bundle, as in the Mallow; or more, as in Hypericum.

Adel'phicus, *Adel'phus*, adelphic.

Adenoca'lyx, (ἀδὴν a gland, καλὺξ the calyx) where the calyx is studded with glandular points.

Adenoph'orus, (ἀδὴν a gland, φέρω to bear) which has glands about it.

Adenophyl'lus, (ἀδὴν a gland, φύλλον a leaf) a leaf studded with glandular spots, or bearing distinct glands.

Adenopo'dus, (ἀδὴν a gland, πõυς a foot) bearing glands on the petioles.

Adenoste'mon, (ἀδὴν a gland, στήμον a stamen) where there are glands on the stamens.

Adflux'ion, (*ad* to, *fluo* to flow) the force by which the sap is drawn towards the leaves; in opposition to the force of propulsion, by which it is propelled forward from the root.

Adglu'tinate, (*ad* to, *glutino* to glue) same as accrete.

Adhæ'rens, adherent.

Adhe'rence, Adhe'sion, (*ad* to, *hæreo* to stick) the complete union, or grafting together of parts, which originally, or in their nascent state, were distinct.

Adhe'rent, Adhe'ring, same as accrete. See adherence.

Adisca'lis, (*a* without, δὶσκος a disk) where the stamens are inserted immediately into the torus, without the intervention of a fleshy disk found in some flowers.

Adminic'ulum, (*adminiculor* to prop) synonyme for fulcrum.

Admoti'vus, (*ad* to, *moveo* to move) in germination, when the episperm investing the extremity of a swollen cotyledon, remains laterally attached to the base of the cotyledon.

Adnas'cens, (*ad* to, *nascor* to be born) synonyme for young bulb; also for suckers of some monocotyledons.

Ad'nate, (*adnascor* to grow to) attached throughout the long length; thus, the anthers are adnate, when their lobes are attached throughout their whole length to the filament; fig. 13; the stipules when they adhere to the peduncles; the bracteæ to the pedicels, &c.

13

Adna'tum, same as *adnascens*.

Adna'tus, adnate.

Adpres'sus, same as *appressus*.

Adscen'dens, same as *ascendens*.

Adventi'tious, (*ad* to, *venio* to come) when some part or organ is developed in an unusual position; as the leaf-buds on various parts of the surface of a stem, instead of being confined, as is generally the case, to the axillæ of the leaves.

Adventitius, adventitious.

Ad'verse, (*ad* towards, *verto* to turn) when one part is placed directly opposite or over against another. Thus, of the anther, when the suture is turned towards the axis or centre of the flower, which is the most usual case. In a curved embryo, where the extremities of the radicle and cotyledons are contiguous, and both turn towards the hilum, they are styled adverse. Where the stigma turns towards the circumference of the flower, so as to face the stamens.

Adver'sus, adverse.

Æqua'lis, *Æ'quans*, equal.

Æquival'vis, equivalvular.

Ae'rial, (*aer* the air) plants or parts of plants which grow entirely above the surface of the earth or water.

Ae'rius, aërial.

Ae'rophyte, (αηρ air, θυτον a plant) a plant which lives entirely out of the ground or water: as many Orchidaceæ, termed Air-plants, whose roots cling to the branches and trunks of trees, and absorb moisture from the atmosphere.

ÆRUGINO'SUS, æruginous.

ÆRU'GINOUS, (*ÆRUGO* verdigris, the green rust of brass) of a rusty colour, whether greenish or reddish-brown.

ÆSCULA'CEÆ, synonym for Hippocastaneæ.

ÆSTIVA'TIO, æstivation.

ÆSTIVA'TION, (*ÆSTIVA* summer quarters) the disposition of the parts of the perianth in the flower-bud. The principal forms of æstivation are the valvular, induplicate, twisted, alternate, quincunxial, vexillary, cochleate, imbricate, calyculate, convolute, and plicate.

ÆTHEOGA'MIC, (ἀήθης unusual, γάμος marriage) a synonym for cryptogamic.

AFFI'NITY, (*AFFINIS* neighbouring) when the relation which plants or groups of plants bear to each other is very close, and depends upon some striking resemblance between their most important organs. Applied in contra-distinction to ANALOGY, where the resemblance, though it may at first appear striking, lies between less important organs. Thus the foliage of the Lathyrus nissolia resembles that of a grass, but there is no affinity between the genus Lathyrus which belongs to the class Dicotyledones, and the grasses which are of the class Monocotyledones.

AGA'MIC, (*a* without, ἀγμος marriage) synonym for cryptogamic.

AGAR'ICOLUS, (*AGARICUS* a genus of fungi, *COLO* to inhabit) applied to some cryptogamic plants which live on agarics.

A'GENUS, (*a* without, γένος offspring) a name applied to cellular acotyledones, which have no distinct increasing surface, but are enlarged by the addition of new parts.

AGGLO'MERATED, (*AGGLOMERO* to crowd together) collected closely together into a head or mass; as the cones on the Scotch-pine, or the flowers of a Scabious.

AGGLOMERA'TUS, agglomerated.

AGGREGA'TED, (*AGGREGO* to assemble) when similar but distinct parts grow crowded together, as the fruit of the mulberry. Much the same as agglomerated.

AGGREGA'TUS, aggregated.

AGRES'TIS, rural. Applied to wi d flowers, whether indigenous or naturalized.

AGYNA'RIUS, *AGY'NICUS*, *A'GYNUS*, (*a* without, γυνὴ a woman) where the pistil is wanting; as in the sterile flowers of Monœcious and Diœcious plants; and also in some double flowers where the stamens and pistils have become petaloid.

AIO'PHYLLUS, (ἀιὼν eternity, φυλλον a leaf) Evergreen.

AIR-CELLS. Cavities in the cellular tissue which are sometimes irregular, but often constructed with great beauty and regularity in the form of hexagonal prisms, &c. They are filled with air, and in aquatics serve the purpose of floating the stem and leaves to the surface of the water. In terrestrial plants they give some stems, as those of rushes (*JUNCI*) a spongy structure.

AKE'NIUM, see *ACHENIUM*.

A'LA, a wing.

ALABAS'TRUS or-*TRUM* the flower-bud.

ALANGIA'CEÆ, (from the genus Alangium) a natural order of Dicotyledones composed of large trees common in the S. of India, and possessing an affinity with Myrtaceæ. It contains only the two genera Alangium and Marlea.

ALA'RIS, (*ALA* a wing) same as axillaris,

ALA'TUS, winged.

ALBES'CENS, albescent.

ALBES'CENT, (*ALBESCO* to grow white) where any colour assumes a pale tinge, or has a hoary appearance.

ALBU'MEN, (*ALBUMEN* the white of an egg) a substance found in many seeds. It is of a farinaceous, oily, or horny consistency, surrounding the embryo wholly or in part, and affording nourishment to the young plant during the earliest stages of germination. Flower obtained from wheat and other corn is composed of it.

ALBUMINO'SUS, containing albumen.

ALBUR'NUM, (*ALBUR'NUM* Sap-wood.) The outermost layers of wood in Exogenous trees, which have not yet passed to the state of Duramen, or Heart-wood.

A'LGÆ, (*A'LGA*, a sea-weed) an order of Acotyledonous plants, of very simple organization, chiefly inhabitants of water, and very many of the sea, (*SEA-WEEDS*); some few are terrestrial, but confined to moist situations. They are very varied in their external appearance; some being composed of homogeneous flattened laminæ, whilst others are capillary, simple or ramified, solid or tubular. Their sporules are either sunk in the substance of the frond, or contained in a peculiar description of tubercles.

ALISMA'CEÆ, (from the genus Alisma) the Water-Plantain Tribe. A natural Order of Monocotyledones, containing only a few aquatic species, with lax tissue, the limb of whose leaves float on the surface of the water. The perianth is distinctly double, the three inner segments petaloid. The stamens and carpels are distinct, and are either six in number or indefinite. The capsules contain one or two seeds, without albumen, and with a curved embryo bent double.

ALKALES'CENT, partaking of the properties of an alkali.

ALLIA'CEOUS, (*AL'LIUM* garlic) possessing the odour of garlic.

ALLIA'CEUS, ALLIA'RIUS, alliaceous.

ALLIGA'TOR, (ALLIGO, to bind, to tie) synonym of *FULCRUM*.

ALLO'CHROUS, (ἀλλὸς another, χρὸα colour) changing from one colour to another.

AL'PINE, (*ALPI'NUS* of the Alps) strictly speaking this term refers to the higher parts of the Alps, in contra-distinction to "Mountainous" (*ALPES'TRIS*,) which designates the middle portions of the higher Alps, or tops of inferior mountains.

ALSINA'CEOUS, (from the genus Alsine) applied to a petal having a short but distinct claw, fig. 14, like those of Alsine.

14

ALTERNANS, alternating.

ALTER'NATE, ALTER'NATING, (*ALTER'NUS* mutual, one after another) when two parts or organs are so placed, that the one is not directly before or over-against the other. Thus when a flower is strictly regular, the parts composing each floral whorl stand opposite the spaces which lie between contiguous parts of the next whorl, fig. 15. In the disposition of the leaves and branches, this term is applied when these organs are apparently disposed without regularity, fig. 16. The æstivation of a perianth is alternate when its parts being disposed alternately in two or more whorls, those which are the outermost also partially overlap those which are within them, fig. 17.

15

16

17

ALTER'NATELY-PINNATE, when the leaflets of a pinnate leaf are not exactly opposite to each other. fig. 18.

18

ALTERNATI'VUS, ALTERNA'TUS, ALTER'NUS, alternate.

ALVEOLA'RIS, ALVEOLA'TUS, alveolate.

ALVE'OLATE, (*ALVE'OLUS* a hollow vessel) studded with cavities, somewhat resembling the cells in a honey-comb; like the receptables of many Compositæ.

Amalthe'a, (ἅμα together, ἀλθέω to heal or increase) name which has been given to an aggregation of dry horny fruit within a persistent calyx which does not become fleshy; as in Agrimonia.

Amarantha'ceæ, (from the genus Amaran'thus) the Amaranth Tribe. An ill-defined order of Dicotyledones composed of humble herbs and a few shrubs. Many are esteemed as potherbs, and some are cultivated for the beauty and durability of their inflorescence, arising from the dense aggregation of their otherwise inconspicuous flowers, and the scarious nature of their deeply coloured bracts or sepals. The sepals are three or five. The stamens are of the same number as the sepals or some multiple of them, distinct or monadelphous. Unless, with Lindley, we exclude the Illecebreæ and some others from the Order, the stamens may be either hypogynous or perigynous, and the perianth either monochlamydeous or dichlamydeous. There is one ovary, with one or few ovules, which becomes a membranous utricle, with the seed pendulous from a central free funiculous. The testa crustaceous, and embryo curved round a central farinaceous albumen (amphitropal.)

Amaryllida'ceæ, (from the genus Amaryllis) the Narcissus tribe. An order of Monocotyledones, in which the perianth is superior, regular and sexpartite; the segments in two whorls, but all are coloured and not separable into distinct calyx and corolla. Six stamens with the filaments sometimes cohering by their dilated bases; and sometimes there is an additional whorl of cohering barren stamens which form a petaloid cup-shaped nectary. Ovary 3-celled, each cell with 1, 2, or most frequently with many seeds. Capsule 3-valved, with loculicidal dehiscence, or else a 1-3-seeded berry. Testa either membranous, brittle, or fleshy. Albumen copious fleshy or corneous, enclosing an erect embryo, which is straight or nearly so. These plants are frequently bulbous, and the flowers subtended by spathaceous bracts.

Ambi'genus, (*ambo* both, *genus* a race or kind) applied to a perianth whose outer surface partakes of the usual character of a calyx, and the inner of a corolla, as in many Monocotyledonous plants.

Ambi'guous, (*ambiguus* doubtful) when certain characters of some part or organ are not well marked, so that it cannot be accurately referred to any well-defined condition. Thus, in an indehiscent pericarp, like that of the Orange, the dissepiments are "ambiguous," because they are equally connected with the axis and paries, and it seems doubtful whether they ought to be considered as expansions of the one or the other. The hilum is "ambiguous," when the seed is so much curved that the apex and base are brought close together. The stipules are so, when they are equally attached to the stem and petiole.

Ambi'parus, (*ambo* both, *paro* to make) producing two kinds; as where a bud contains both flowers and leaves.

Ambleocar'pus, (ἀμϐλόω to be abortive, χαρπὸς fruit) where several ovules being abortive a few only become perfect seeds.

Ambrosi'acus, (ἀμϐρόσιος divine, sweet) possessing a strong scent like Ambrosia.

Am'ent, *Amen'tum*, a catkin.

Amenta'ceous, *Amenta'ceus*, having the form of a catkin.

Ammo'philus, (ἄμμος sand, φίλεω to love) growing spontaneously in sandy soil.

Am'nios, (*amnios;* ἀμνίος a fœtal membrane) a viscous fluid which in some ovules surrounds the embryo in its earliest state, and a portion of which afterwards thickens into the "perisperm."

Amor'phous, *Amorphus* (α without, μορφὴ form) where the form is not well defined, or distinct.

Amphan'thium, (ἀμφὶ around ἄνθος a flower) a dilated receptacle on which numerous florets are seated, as in the Compositæ, Ficus, &c.

Amphi'bious, *Amphi'bius* (ἄμφω both βίος life) growing equally in water and on dry land.

Amphicar'pic, *Amphica'rpus;* (ἄμφω both, καρπὸς fruit) possessing fruit of two kinds—either as regards its form, or period of ripening.

Amphigas'trium, (ἀμφὶ about, γαστὴρ the belly) a stipular appendage peculiar to certain Jungermanniæ, which clasps and covers their stems.

Amphisar'ca, (ἀμφὶ about, σαρξ flesh) an indehiscent multilocular fruit, dry externally and pulpy within.

Amphisper'mium, (ἀμφὶ about, σπέρμα a seed) where the pericarp so closely invests the seed as to maintain the same form with it.

Amphi'tropal, *Amphi'tropus* (ἀμφὶ about τρέπω or τρόπέω to turn) when the embryo is so much curved, that both ends are brought close together and turned towards the hilum. fig. 19.

Am'phora. This term has been applied to the lower division of the peculiar form of capsule styled a "pyxidium."

Amplec'tans, Amplecti'vus, Amplex'ans, embracing.

Amplex'icaul, *Amplex'icaulis,* (*amplector* to embrace, *caulis* the stem) when the peduncle, leaf, or stipule is dilated at the base, and extends partially round the stem, so as to clasp it. fig. 20.

Ampu'lla, a Bladder.

Ampulla'ceous, *Ampulla'ceus, ampullæformis;* (*Ampulla,* a vessel swollen out in the middle) where some part, as the monopetalous corolla of certain heaths (*Erica,*) is swollen out like a bottle fig. 21.

Amy'gdala, a kernel.

Amyla'ceous, *Amyla'ceus;* (ἄμυλον flour) of the nature of fecula.

A'nabix, pl. *ices* (ἀναϐἰοω, to revive) those parts of the nutritive organs of some cryptogamic plants which are contiuually perishing below, but vegetate above, as the stems of Lycopodium, fronds of Jungermannia, &c.

Anandra'rius, (*a* without, ἀνὴρ a man) a double flower in which the stamens have entirely disappeared.

Ana'nthus, (*a* without, ἄνθος a flower) flowerless.

Anasa'rca, (*ανα* through, σαρζ flesh) a disease in plants termed dropsy; arising from a superabundance of fluids in their tissue.

Anastomo'sis, (*Anastomosis;* ἀναςομωσις passage of one vein into another) where branches of vascular tissue reunite; as in the reticulations formed by the nerves or veins of many leaves.

Ana'tropous, *Anatropus;*(*ανα* over, τρεπω to turn) where the chalaze (c) is distant from the hilum (h) and the apex (a) of the ovule appears completely reversed and brought near the hilum, fig. 22, as in Liliaceæ.

Anci'pital, (*Anceps* two-edged) flattened or compressed, with two edges more or less sharp; as the stems of Sisyrinchium anceps.

ANDROCŒ'UM, (ἀνὴρ ἀνδρὸς a man) the stamens taken collectively; just as corolla signifies the aggregate of the petals, and calyx of the sepals.

ANDRODY'NAMUS, (ἀνὴρ a man, δυναμις power) a term which has been employed for Dicotyledonous plants; in which Class the stamens and petals are generally highly developed.

ANDROGYNA'RIS (ἀνὴρ a man, γυνὴ a woman) double flowers in which both stamens and pistils have become petaloid.

ANDRO'GYNOUS, *ANDROGYNUS*: (ἀνὴρ a man, γυνὴ a woman) this term when applied to a plant is used as a synonym for Monœccous, where some of the flowers of an individual contain stamens only and others pistils only; but applied to a flower it is synonymous with hermaphrodite, where both stamens and pistils are present within the same perianth.

ANDROPETALA'RIUS, (ἀνὴρ a man, πέταλον a petal) double flowers in which the stamens have become petaloid, the pistils remaining unchanged.

ANDRO'PHORUM, (ἀνὴρ a man, φέρω to bear) whenever the filaments are united together so as to appear as a single support to several anthers, as in monadelphous plants, &c.

ANFRA'CTOUS, *ANFRACTUO'SUS*, (*ANFRA'CTUS*, a turning or winding) presenting sinuosities; as the anthers of gourds and cucumbers, fig. 24. Also a synonym for spirally twisted.

24

ANGIOCA'RPUS, (ἀγγεῖον a vessel or receptacle, καρπος fruit) where the fruit is invested by some extraneous organ so as not to be distinguishable at first sight, as in Coniferæ and Cupuliferæ.

ANGIOSPE'RMIA, (ἀγγεῖον a vessel, σπέρμα seed) the second Order of the 14th Class of the Linnean system, *DIDYNAMIA*; characterized by the fruit being capsular, fig. 25, and not composed of four nuts as in the first order *GYMNOSPERMIA*.

25

ANGIOSPER'MUS, fruit formed as in Angiospermia.

ANGLE, (*A'NGULUS*, an angle) not limited in botany to the inclination of two lines, but is often used to express the inclination of two planes forming an edge, as in "angular stems."

A'NGULAR, (*A'NGULUS*, an angle) where an organ offers a determinate number of angles, as the quadrangular stems of Labiatæ.

ANGULA'RIS, belonging or attached to an angle or edge; as

where pubescence occurs on the edges of the stems of Cacti.

ANGULA'TUS, angular.

ANGU'LINERVED, *ANGULINE'RVIUS*, (*A'NGULUS* an angle, *NERVUS* a nerve) where the nerves or veins in the limb of a leaf branch off from the end of the petiole, or from each side of a mid-rib at acute angles, and then become more or less sub-divided and ramified, and assume the reticulated appearance characteristic of the leaves of a great portion of Dicotyledons.

ANGULO'SUS, angular.

ANGUSTIFO'LIUS, (*ANGUSTUS* narrow, *FOLIUM* a leaf) where the breadth of a leaf is small when compared with its length. Ex. Epilobium angustifolium.

ANGU'STISEPTUS, (*ANGU'STUS* narrow, *SEPTUM* division) where the dissepiment (d) between the two cells, is very narrow compared with the whole breadth of the silicula. Applied principally in the Cruciferæ. fig. 27.

ANISA'TUS, (*ANI'SUM* anise) partaking of the scent of anise.

ANISO'BRIOUS, *ANISOBRIA'TUS* (ἄνισος unequal, ἔμβρυον an embryo) a name which has been given to the embryo of monocotyledons, because one side is supposed to possess a greater developing force than the other; by which means a cotyledon is formed on that side, whilst none arises on the other.

ANISODY'NAMOUS, *ANISODY'NAMUS*, (ἄνισος unequal, δυναμις force) synonym for Aniso'brious.

ANISOME'RICUS, (ἄνισος unequal, μὲρος a part) where the corresponding parts of a flower are not all regular, or alike.

ANISOPE'TALUS, (ἄνισος unequal, πὲταλον a petal) where the petals are of unequal size. Ex. Prangos anisopetalus.

ANISOPHY'LLUS, (ἄνισος unequal, φυλλον a leaf) where one of two leaves placed oppositely is much larger than the other. Ex. Ruellia anisophylla.

ANISOSTE'MONOUS, *ANISOSTEMONIS*, (ἀνισος unequal, ςήμων a stamen) where the number of stamens in a whorl is different from the number of parts in a whorl of the perianth. Thus in Scabiosa and many other Dipsaceæ, the corolla is formed by the adhesion of five petals, whilst there are only four stamens.

ANISOSTEMOPE'TALUS, synonym for *ANISOSTE'MONUS*.

ANNEX'US. Adnate.

AN'NUAL, *ANNO'TINUS*, *ANNUA'LIS*, (*ANNUUS*, yearly) applied to a plant, signifies that it produces seed and dies within the same year in which it first germinated. The symbol (·) or (I) is used to denote this. An annual leaf is one which falls in the autumn, as contradistinguished from an evergreen which lasts through the winter.

AN'NULAR, *ANNULA'RIS*, *ANNULA'RIUS*, (*AN'NULUS* a ring) applied to any organ or set of organs disposed in a circle, and resembling a ring. In the vascular tissue, an "annular-vessel" is a cylindrical membranous tube marked at intervals with transverse stripes or rings, fig. 28, probably composed of fibre, similar to that of which the tracheæ are formed, of which vessels these are generally considered to be a modification.

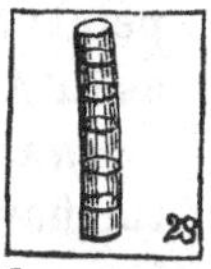

28

ANNULA'TUS, ringed.

AN'NULUS, a ring.

ANO'MALOUS, *ANO'MALUS*, (*α* not, ὁμαλὸς equal) where a plant is very unlike the great majority of those to which it is most nearly allied. Or where some organ is remarkable for the singularity of its shape, which cannot readily be assimilated to any common object for the purpose of comparison; as in the petals of Delphinium and Aconitum.

ANONA'CEÆ, (from the genus ANO'NA) a small natural order of Dicotyledons.

ANTE'RIOR, *ANTE'RIOR*, (*ANTE* before) refers both to time and position. In the latter application, those parts are anterior which are placed in front of others, or outwardly with respect to the axis about which they are arranged. This term has been extended to signify the direction in which an organ is turned; and in the flowers of many Orchideæ, which are inverted by a twist in the germen, the stigma has been termed anterior, because it then faces the outward, though naturally posterior portions of the perianth.

AN'THER, *ANTHE'RA*, (ἀνθηρὸς flowery) that portion of the stamen which contains the pollen. It is most frequently formed of two distinct cells, and is very variously shaped, and generally attached towards the summit of a filament, though it is sometimes sessile or without one. When the attachment is at its base, the anther is said to be terminal, fig. 29, *a*, when by the middle of the back it is horizontal, &c. *b*. In some cases, as in

a b 29

Epacrideæ, there is only one cell. The cells usually burst by a longitudinal slit to emit the pollen, but in some cases this escapes through pores only.

ANTHERI'FEROUS, *ANTHERI'FERUS*, (*ANTHE'RA* an anther, *FERO* to bear) a part which supports the anthers; as the filament generally, or the body formed by the union of the filaments in adephic flowers.

ANTHERO'GENOUS, *ANTHERO'GENUS*, (ἀνθηρα the anther, γεννάω to beget) those double flowers whose supernumerary petals have originated from the transformation of anthers; as in Aquilegia vulgaris, variety corniculata.

AN'THESIS, (ἀνθέω to flower) either signifies the time when the flower has arrived at perfection and the anther is just bursting; or is used to express the phenomena themselves, exhibited upon the expansion of the flower. Sometimes it is restricted to the mere bursting of the anthers.

ANTHO'DIUM, (ἀνθος a flower, δύω to put on.) By some restricted to the involucrum of Compositæ, by others extended to signify the whole capitulum in this order.

ANTHO'PHORUS, (ἀνθος a flower, φέρω to carry) when the receptacle or torus is lengthened within the calyx, the part which supports the inner portions of the flower is thus termed, as in Silene; see fig. 30.

ANTICLINAN'THUS, (ἀντὶ before, κλίνη a bed, ἀνθος a flower) the inferior and scaly part of some receptacles (*CLINANTHUS*) in the Compositæ.

ANTI'CUS, (the fore-part) synonym for *INTRORSUS*.

ANTI'TROPAL, *ANTI'TROPUS*, (ἀντὶ opposite, τρέπω to turn) where the embryo lies reversed with respect to the seed—its cotyledons, or upper extremity, being directed towards the hilum, or base of the seed, as in Daphne.

ANTROR'SUM, (*ANTE* before) having an upward direction towards the summit of some part.

ANTRUM, (a cave) has been used as a synonym for the fruit termed *POMUM*.

APERTUS, naked.

APE'TALOUS, *APE'TALUS*, (α without, πέταλον a petal) a flower destitute of true corolla, however much the calyx may be coloured and petaloid.

APEX, (*APEX* the top of any thing) formerly used as a synonym for the anther. Applied to the opposite extremity of any organ to that by which it is attached, and which is considered its base. The apex may occasionally be brought close to the base, as the apex of the nucleus in the case of anatropous and campulitropous seeds.

APHYL'LOUS, *APHYL'LUS*, (α without, φυλλον a leaf) destitute of leaves. Sometimes signifies their partial or imperfect production.

API'CILLARY, *APICILLA'RIS*, (*APEX* a summit) applied to any organ which is inserted upon or towards the summit of another; or used to signify some circumstance connected with the summit of an organ; as in the Caryophyllaceæ, where the dehiscence of the capsule is at the summit.

API'CULATE, *APICULA'TUS*, furnished with an apicula. Pointletted.

API'CULA, *API'CULUS*, (*APEX* a sharp point) a sharp but short point in which a leaf or other organ terminates, and which is not very stiff.

APOCAR'POUS, *APOCAR'PUS*, (ἄπο from, apart, καρπος fruit) properly signifies where the carpels are quite free from adhesion; as in Ranunculus, Caltha, &c. but is also applied where they are merely united in so partial a manner, that the compound pistil is distinctly separable into them.

APO'PHYSIS, *APO'PHYSIS*, (ἀποφύω to spring from) any irregular swelling on the surface. More especially applied to an inflation of the fruit-stalk immediately below the theca of some mosses.

APOTHE'CIUM, (ἀπο upon, θήκη a chest) the organ of fructification peculiar to Lichens, which contains their sporules, and is frequently cupshaped.

APPEN'DAGE, *APPEN'DIX*, a part superadded to another; thus the leaves are appendages to the stem.

APPEN'DENT, *APPEN'DENS*, (*APPENDO* to hang by) when the hilum is towards the upper part of the seed, which is sessile, or nearly so, on the placenta; as in the plum and other stone-fruit.

APPENDICULA'TUS, furnished with any kind of "appendage."

APPLE. This term is extended beyond its familiar signification, the fruit of the apple-tree, to other fruit constructed on the same plan, as those of the Medlar, fig. 31, Hawthorn, &c. It is formed out of two or more concrete inferior carpels, which have their paries membranous or bony; they are closely invested by

31

the tube of the calyx, which unites with their substance and becomes pulpy. The carpels ultimately appear as so many cells in the midst of this succulent pulpy mass.

Applicati'vus, *Applica'tus*, (applied, laid to) where two surfaces are applied face to face without folds of any sort; as in the venation of the leaves of Aloe, exhibited by the transverse section.

Apposi'te, (*appo'situs* laid near, set upon) where similar parts are similarly placed. Thus, in most anthers the dehiscence of each lobe is towards the same side. Two ovules in the same cell are apposite when they are attached close to the same point of the placenta.

Appress'ed, (*appres'sus* pressed hard together) where a part lies close to another throughout its whole length; as the pubescence on some leaves and branches. Branches which stand parallel and close to the stem, as in the Lombardy poplar, are termed appressed; and the stem itself is said to be so when it trails close along the ground.

Approx'imate, *Approxima'tus*, (*ad* to, *pro'ximo* to approach) parts which are close together, but not united.

Ap'terous, *Ap'terus*, (*α* without, πτερὸν a wing) without any membranous appendage like those on certain petioles, seeds, &c. and which are termed wings.

Apyre'nus, (*α* without, πυρὴν a seed) fruit which produces no seeds, as many cultivated varieties of Pine-apple, Orange, &c.

Aqua'tic, (*aqua'ticus*, *aqua'tilis* living or growing in water) applied to all plants which grow in the water, whether they are entirely submersed as the Confervæ, float on its surface as the Lemnæ, or, having the roots fixed in the soil, raise their leaves and flower above the water as the Water-lilies (Nymphæa).

A'queous, *A'queus*, (*aquo'sus* watery, or resembling water) generally indicates some nearly colourless tint.

Aquifolia'ceæ, (from the old genus Aquifolium, now Ilex) the Holly tribe; formerly classed with Rhamneæ, and now considered as a tribe of Celastrineæ by some, and by others as a distinct order of Dicotyledons.

Aquilari'neæ, or Aquilaria'ceæ, (from the genus Aquilaria) the Agallochum tribe. A small natural order of monochlamydeous Dicotyledons.

ARA'CEÆ, or ARO'IDEÆ, (from the genus ARUM) a natural order of Monocotyledons.

ARACHNO'ID, *ARACHNOI'DEUS*, (ἀράχνης a spider) composed of soft downy fibres, resembling the web of the gossamer spider; as the pubescence on the leaves of Sempervivum arachnoideum, Calceolaria arachnoidea, &c.

ARA'NEUS, *ARANEO'SUS*, (*ARA'NEA* a spider) arachnoid.

ARALI'ACEÆ, (from the genus ARALIA) a natural order of Dicotyledons.

ARBORE'SCENT, *ARBORE'SCENS*, *ARBO'REUS*, (*ARBORES'CO* to grow to the size of a tree, and *ARBO'REUS* having the form of a tree.) The terms are used indiscriminately to signify any plant which attains the size, or assumes the form of a tree.

ARBUS'CULA, a small shrub with the appearance of a tree, like many heaths (*ERICEÆ.*)

ARBUSCULA'RIS, ramified like a little tree.

ARCU'ATUS, bent like a bow, so as to form a large arc of a circle. Curved.

ARDISI'ACEÆ, (from the genus ARDISIA) a synonym for Myrsinaceæ.

ARE'OLÆ, (little beds in a garden) spaces distinctly marked out on a surface. Otherwise used synonymously with small cells or cavities.

ARE'OLATE, *AREOLA'TUS*, covered with areolæ.

ARGENTE'US, (of silver or of the lustre of silver) white with a tinge of grey, and glittering with somewhat of a metallic lustre; as the silky hair on the leaves of Evolvulus argenteus.

ARHI'ZUS, (*a* without, ῥίζα a root) a term which has been applied to plants whose roots are very small. Also to those whose reproductive organs have no radicle, and consequently no true embryo.

ARILLA'TUS, furnished with an aril.

ARI'L, *ARIL'LUS*, an expansion of the funicular chord, rising round certain seeds in the form of an integument, generally more or less fleshy, as in the genus Euonymus, fig. 32. The Mace of commerce is the arillus of the Nutmeg.

32

ARIS'TA, (the beard of corn) an awn.

ARISTA'TUS, (furnished with an *ARISTA*) awned.

ARISTOLO'CHIÆ, or ARISTOLOCHIA'CEÆ, (from the genus ARISTOLOCHIA) the Birth-wort tribe. A natural order of Dicotyledons.

Aristula'tus, furnished with a very small arista.

Ar'ma, (all kinds of armour) such appendages as prickles, thorns, &c. which serve as defences to protect plants against the attacks of animals.

Armeni'acus, (*armeni'acum* fruit like an apricot) of an apricot colour, yellow tinged with red.

Aroi'deæ, synonym for Araceæ.

Arrec'tus, erect.

Arrow-shaped, or -headed, synonym for Sagittate.

Articula'tion, (*Arti'culus* a knot or joint) a place where a discontinuity of tissue naturally takes place, without the appearance of its having been torn asunder; as where many leaves fall off in autumn. The joints of stems are sometimes termed articulations, even when no such separation takes place. The transverse diaphragms which indicate the separation of distinct cells in tubular-shaped Algæ have also been called articulations.

Arti'culate, -ed, *Articula'tus*, furnished with articulations.

Artifi'cial, *Artificia'lis*, applied to some characteristic by which a plant may be distinguished or separated from others, without reference to those other circumstances by which its affinities are established.

Artiphyl'lous, *Artiphyl'lus*, (ἄρθρον a joint, φυλλον a leaf) where a joint of the stem bears leaf-buds.

Ascel'lus, synonym for *ascus*.

Ascen'ding, Ascen'dent, (*ascendens* ascending) where an organ starting horizontally or rising obliquely from the base curves upwards, and ultimately attains a vertical position; as in many stems, fig. 33, *a*; and in seeds fixed towards the base of the pericarp, *b*. It is applied generally in opposition to descending.

As'cus, (ἀσκὸς a leather bottle) a membranous tubular cell, of which several are sunk in the substance of Lichens and Fungi containing their sporules.

Asci'dium, (ἀσκι'διον a small bottle) an appendage termed a pitcher.

Asclepia'deæ, or Asclepiada'ceæ, (from the genus Asclepias) an extensive order of Dicotyledons.

Asimi'na, synonym for *syncarpium*.

Aspara'geæ, or Asparagi'neæ, (from the genus Asparagus) a group of Monocotyledons, sometimes considered as a distinct order, but which may be referred (as a subordinate section) to Liliaceæ.

ASPERGILLIFOR'MIS, (*ASPERGIL'LUM* a brush used to sprinkle holy water, *FORMA* shape) little tufts of hair which collected together assume the form of a brush; as those on the stigma of Arundo phragmites.

ASPHODE'LEÆ, (from the genus *ASPHODELUS*) a group of Monocotyledons, which may be considered as a distinct natural order, or referred to Liliaceæ as a subordinate tribe.

ASSIMILA'TION, *ASSIMILA'TIO*, (*ASSI'MILO* to assimilate) that act by which a plant (or other organized being) converts nutritious matter into its own substance.

ASSUR'GENS, (*ASSURGO* to ascend) synonym for *ASCENDENS*.

ATAVIS'M, *ATAVIS'MUS*, (*ATAVUS* an ancestor) the resemblance borne by a plant to an original race or stock from which it is descended, though it may have sprung from the seed of some different variety of the same species.

A'TER, (in composition ATRO-) pure black, as distinguished from Niger.

ATHEROSPERMA'CEÆ, or ATHEROSPER'MEÆ, (from the genus ATHEROSPERMA) a small natural order of Dicotyledons, closely allied to Monimiaceæ.

ATTENUA'TED, (*ATTENUA'TUS* diminished) where the breadth is gradually diminished towards either extremity; as the bases of some leaves.

AUC'TUS, (increased) where some parts appear to be superadded to an organ beyond what is usual. Thus the calyx in Dianthus has a distinct whorl of bracteal scales at its base.

AURANTI'ACUS, *AURAN'TIUS*, (*AURANTIUM* an orange) of an orange colour.

AURANTIA'CEÆ, (from the Citrus aurantium) the orange tribe. A natural order of Dicotyledons.

AURA'TUS, *AU'REUS*, (golden) of a bright golden colour; composed of yellow with a small portion of red.

AURI'CULA, (the ear, or lap of the ear) rounded appendages at the base of some leaves; as in those of Salvia officinalis; or those otherwise called wings on the petioles, as in the Orange, fig. 34. The stipules of Jungermanniæ have also received this name.

34

AURICULA'TE, *AURICULA'TUS*, (having auriculæ) eared.

AUSTRA'LIS, (southern) frequently applied to plants which grow in warm climates, without regard to their being strictly confined to the southern hemisphere.

Autocar'pian, *Autocarpea'nus*, (ἀυτὸς alone, καρπος fruit) synonym for a "superior" fruit, one which contracts no adhesion with the perianth.

Avena'ceous, (*avena'ceus* of oats) bearing some relation, or casual resemblance to oats.

Ave'nius, (*a* without, *vena* a vein) veinless.

Aver'sus, (turned back) inverse.

Awl-shaped, narrow and gradually tapering to a fine point.

Awn, a stiff bristle-like appendage to various foliaceous and floral organs, especially the glumes and paleæ of grasses. In some cases it is evidently a prolongation of the mid-nerve.

Awned, furnished with an awn.

Axe-shaped, having somewhat the form of an axe, as the fleshy leaves of some Mesembryanthema.

Axil, (*axilla* the arm-pit) the upper angle formed by the attachment of a leaf or branch to its support.

Axil'lary, *Axilla'ris*, (*axilla* the arm-pit) occurring in an axil; as the young buds in the axils of the leaves of most plants, fig. 35.

35

Axis, (*axis* an axle tree) an imaginary line forming a centre round which an organ is developed. The term is also extended to signify the organ round which others are attached; thus the stem is termed the ascending axis, and the root the descending axis of vegetation.

Azure, *Azu'reus*, of a lively pale blue, like the sky.

Bac'ca, a berry.

Baccaula'rius, (*bacca* a berry) synonym for *carcerulus*.

Baccate, *Bacca'tus*, *Bacci'ferus*, (*bacca* a berry) bearing berries; or having a succulent nature like that of berries, expressed also by *Bacciformis*.

Bacil'la, *Bacil'lus*, a small bulb.

Back, the side which is turned from the part to which any organ is attached.

Ba'dius, a reddish or chestnut brown.

Balaus'ta, (βαλαύστιον pomegranate flower) fruit formed like the pomegranate—indehiscent, inferior, with many cells and seeds. The seeds coated with pulp, fig. 36.

36

Bald, destitute of pubescence, or downy appendages.

Banded, when stripes of colour are arranged transversely.

Band-shaped, a variety of "linear," where the length is considerable, as in the leaves of Zostera marina.

Barb, a double hook at the end of some bristles; as on the fruit of Echinospermum lappula.

Barba'tus, bearded.

Bark, the external coating of the stems and roots of phanerogamous plants,—more distinctly formed on Dicotyledones than on Monocotyledones; and in trees of the former class, is annually increased in thickness by the addition of a fresh layer.

Ba'sal, ***Basila'ris***, (*basis* the base) attached to the base of any organ or part.

Base, ***Ba'sis***, that extremity at which an organ is attached to its support, and by which the nourishing vessels enter it.

Basigy'nium, (βάσις a base, γυνη a woman) synonym for Thecaphore.

Basiner'vis, (*basis* the base, *nervus* a nerve) where the nerves of a leaf, as in the grasses, proceed from the base to the apex without subdividing.

Beaked, terminated by a long pointed projection, as the pod of the Radish.

Beard, a synonym for Awn.

Bearded, when tufts of hair-like pubescence are attached to various parts of a plant, as those on the lip of Chelone barbata; and those on the leaves of Mesembryanthemum barbatum.

Bedeguar, a tumour or excrescence on the branches and leaves of roses, coated with fibrous expansions of the tissue. This appearance is occasioned by the puncture of a Cynips.

Bell-shaped, having a tubular and inflated form, so as to resemble a bell, as the corolla of many Campanulæ.

Bellying, swelling out on one side, as the tube of the corolla in many plants of the order Labiatæ.

Berried, synonym for Baccate.

Bi-acu'minate, ***Biacumina'tus***, (*bis* twice, *acuminatus* pointed) where there are points in two directions; as in the pubescence on the leaves of Malpighiaceæ, where it is attached by the middle.

Biconjuga'to-pinna'tus, (*bis* twice, *conjungo* to join together, *pinnatus*, feathered) where two secondary petioles meet at the apex of a general petiole, and each bears leaflets arranged in a pinnate form, fig. 37.

37

BICONJUGATE, *BICONJUGATUS*, (Bis twice, *CONJUNGO* to join together) where two secondaryp etiolesstand at the apex of a general petiole, and each bears a single pair of leaflets,

BICOR'MIS, (*BIS* twice, *CORNU* a horn) furnished with two pointed appendages resembling horns; as the anthers of some Ericacæ, &c.

BIDEN'TATE, *BIDENTA'TUS*, (*BIS* twice, *DENTATUS* toothed) when the teeth forming the marginal incisions of leaves, are themselves edged by smaller teeth; also when the divisions of some part are limited to two in number, as the leaves of Cambessedia bidentata.

BIDIGITA'TO-PINNA'TUS synonym for *BICONJUGATO-PINNATUS.*

BIDIGITA'TUS, synonym for Biconjugatus.

BI'DUUS, (*BIDUUM* two days long) lasting for two days.

BIEN'NIAL, (*BIENNIS* of two years' continuance.) A plant which produces only leaves during the first year of its growth; and in the second bears seed, and then dies.

BIFA'RIUS, (two manner of ways) synonym for Distichus.

BI'FID, *BIFI'DUS*, divided about half way to the base into two parts: as the petals of Erophila vulgaris.

BIFO'LIOLATE, *BIFOLIOLATUS*, (*BIS* twice, *FOLIOLA* leaflet) synonym for binate.

BIGEM'INATE, *BIGEMINA'TUS*, (*BIS* twice, and GEMINATUS doubled) synonym for biconjugate.

BIGUGA'TUS, (*BIGUGUS* yoked or coupled) where there are two pairs of leaflets on a pinnate leaf, with or without a terminal leaflet, fig. 38.

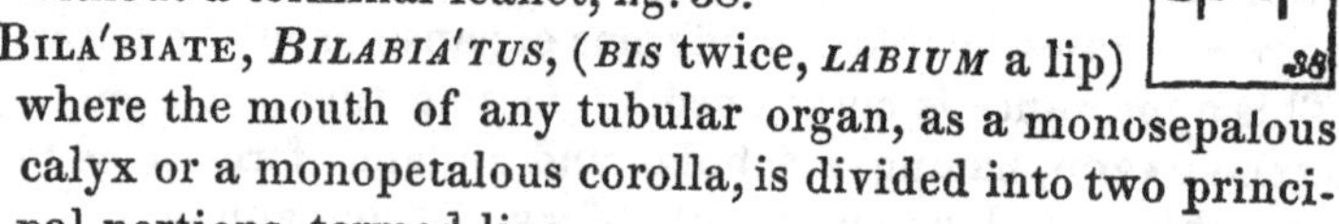

BILA'BIATE, *BILABIA'TUS*, (*BIS* twice, *LABIUM* a lip) where the mouth of any tubular organ, as a monosepalous calyx or a monopetalous corolla, is divided into two principal portions, termed lips.

BILA'TERAL *BILATERA'LIS* (*BIS* twice, *LATUS* a side) arranged on or towards opposite sides, as the leaflets of Taxus baccata.

BILO'BED, *BILOBA'TUS*, *BILO'BUS*, (*BIS* twice, *LOBA* a lobe) divided into two lobes, as the anthers of most flowers; the embryo of dicotyledones, &c.

BILO'CULAR, *BILOCULA'RIS*,(*BIS* twice, loculus a cell) containing two cavities; as the two-celled fruit of many plants; for example, the berry of Ligustrum vulgare.

BINA'TE, *BINA'TUS*, (Binus, by couples) where a leaf is composed of two leaflets placed at the extremity of a common petiole, as in Zygophyllum fabago. The term is sometimes extended to a simple leaf, which is nearly divided into two equal parts.

BIPAL'MATE, *BIPALMA'TUS*, (*BIS* twice, PALMA the palm) when the leaflets are arranged in a palmate manner upon secondary petioles similarly arranged with regard to the primary petiole.

BIPARTI'BILIS, (*BIS* twice, *PARS* a part) capaple of being readily divided into two similar parts, as the fruit of the umbelliferæ.

BIPAR'TITE, *BIBARTI'TUS*, (*BIS* twice, *PARS* a part) deeply divided into two parts; the incision extending beyond the middle, or within about two-thirds, of the distance towards the base.

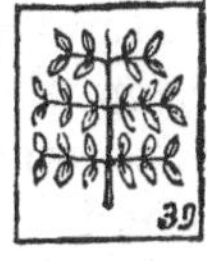

BIPINNA'TE, *BIPINNA'TUS*, (*BIS* twice, *PINNA* a feather) where the leaflets on the secondary petioles of a doubly compound leaf, are arranged in a pinnate manner; the secondary petioles themselves being similarly disposed on the primary, fig. 39.

BIPINNA'TIFID, *BIPINNATIF'IDUS*, (*BIS* twice, *PINNA* a feather, *FINDO* to divide) where the divisions of a pinnatifid leaf are themselves divided in a similar manner.

BI'PLICATE, *BIPLICA'TUS* (*BIS* twice, *PLICO* to fold) doubly to fold) doubly folded in a transverse manner; as in the section of some cotyledons.

BISER'RATE, *BISERRA'TUS* (*BIS* twice, *SERRA'TUS* sawed) where the serratures of a leaf are themselves serrated.

BITER'NATE, *BITERNA'TUS*, (*BIS* twice, *TERNUS* three and three together) where the leaflets of a doubly compound leaf are arranged in a ternate manner, on secondary petioles, similarly arranged as the primary.

BITTEN, where some organ terminates abruptly, (or is truncated) and the end seems to be irregularly torn, as if it were bitten off, as in the leaf of Caryota urens, the root of Scabiosa incisa.

BLADDERS, hollow membranous appendages on the roots of Utriculariæ, which are filled with air, and cause these plants to float; also cellular expansions of the substance of many Algæ, which are filled with air.

BLADDERY, when a tubular organ, as the calyx of Silene inflata, is thin, membranous, and swollen.

BLADE, synonym for the limb of a leaf.

Blaste'ma, (βλαστανω to germinate) the whole of the Embryo after the cotyledons have been abstracted.

Blastus, (βλαστανω to germinate) a name which has been given to the peculiar form assumed by the plumule in the embryo of the Graminaceæ.

Blotched, where colour is irregularly disposed in broad patches.

Blunt, terminating in a rounded manner, without tapering to a point, or without appearing to be abruptly cut off.

Boat-shaped, short, concave, and keeled as the glumes of Phalaris Canariensis.

Bony, where the texture is close, and the substance hard and brittle, as in the stones of peaches, plums, &c.

Borigi'naceæ, (from the genus Borago) the Borage tribe. A natural order of Dicotyledones, containing only the genus from which it is named.

Bordered, when the margin is characterized by a distinction in colour, texture, or other consideration from the rest of any part.

Bossed, where a rounded surface has a projecting point in the centre.

Bolulifor'mis, sausage-shaped.

Brachia'te, *Brachia'tus*, (*brachium* the arm) where successive opposite pairs of branches are placed at right angles to each other, as in the Ash, fig. 40.

Bra'chium, synonym for ulna, an ell.

Bract, *Brac'tea*, (*bractea* a thin leaf) the leaves more or less modified in form, which are seated on the peduncles and pedicels. They are frequently reduced to mere scales, and are sometimes highly coloured and resemble the parts of a perianth.

Bractea'tus, either applied to a plant which possesses bracts; or to one whose bracts are remarkable for their size and form.

Bracte'ola, (diminutive of *bractea*) small bracts, seated on the pedicel, are sometimes distinguished from the rest by this term.

Branch, the developed state of a leaf bud, when similar to the main stem or trunk. Though branches are usually considered to be subdivisions of the trunk itself, they more closely resemble an aggregation of separate individuals grafted upon it.

BREXIA'CEÆ, (from the genus Brexia) a natural order of Dicotyledones, containing only the genus from which it is named.

BRISTLE, short or stiff hair, like the pubescence on Echium vulgare.

BRISTLE-POINTED, terminating very gradually in a fine point, like the leaves of many mosses.

BROMELIA'CEÆ, BROMEL'LIÆ, (from the genus Bromelia.) the Pine Apple tribe. A natural order of Monocotyledones.

BRUNIA'CEÆ, (from the Genus Brunia) a natural order of Dicotyledones.

BRUN'NEUS, deep brown, formed by mixing dark grey with red.

BRUNONIA'CEÆ, (from the Genus Bruno'nia) this natural order of Dicotyledones has been formed by Lindley to admit the single genus Brunonia, heretofore classed under Goodenoviæ.

BRUSH-SHAPED, when hair is collected round the extremity of a slender organ, so as to resemble a bottle brush.

BUCKLER-SHAPED, formed like a round buckler, with a thickened or elevated rim.

BUD, either the nascent state of a branch, when it is termed a leaf-bud; or of the inflorescence, when it is called a flower bud.

BULB, *BULBUS*, a modified form of the Leaf-bud, in which the subordinate parts are more or less fleshy. It is generally underground and seated immediately over the neck of the root. The two principle forms are, 1st, the scaly bulb, fig. 41, where the modified leaves assume the character of fleshy scales, as in the white lily; and 2nd, the laminated or tunicated bulb, where the leaves form successive coats one over the other, the outermost becoming more or less membranaceous, as the onion.

41

BULBIF'EROUS, *BULBIF'ERUS* (*BULBUS* a bulb, *FERO* to bear) plants, or the parts of them, which produce bulbs.

BULBIL'LUS (diminutive for *BULBUS*) applied more especially to those aerial buds, on the stem or elsewhere, which occasionally assume the character of bulbs, as in the Lilium tigrinum.

BULBO-TUBER, synonyme for Cormus.

BUL'BULUS, (diminutive for *BULBUS*) a young bulb which

originates on the old one, from whence the plant sprung.

BUL'BUS, a bulb.

BULLA'TUS, (*BULLA* a bubble) where the spaces between the nerves of a leaf present convexities on one side, and concavities on the other, giving the whole surface a blistered appearance; as in Ranunculus bullatus.

BURMANNIA'CEÆ, (from the genus Burmannia) a natural order of Monocotyledones.

BURSERA'CEÆ, (from the genus Bursera) may be considered either a distinct natural order, or a tribe of Terebintaceæ.

BURSIC'ULA, (*BURSA* a purse) a membranous sack, single or double, at the base of the anther in some Orchidaceæ, in which the retinaculum, or glandular base of a pollen-mass is lodged. *FIG.* 42.

42

BUSH, a low shrub, densely branched from the very surface of the ground.

BUTOMA'CEÆ BUTO'MEÆ, (from the genus Butomus) the Flowering-rush Tribe. Either a section of Alismaceæ, or a distinct order of Monocotyledones.

BUTTERFLY-SHAPED, Synonyme for Papilionaceous.

BYSSA'CEOUS, *BYSSA'CEUS* (*BYSSUS*, fine flax) composed of delicate filaments, resembling cotton or wool; as the roots of many Agarics.

BYTTNERIA'CEÆ, (from the genus Byttnera) either a section of Sterculiaceæ, or a distinct order of Dicotyledones.

CAC'TI, CACTIA'CEÆ (from the genus Cactus) the Indian-fig Tribe. A natural order of Dicotyledones.

CADU'COUS, (*CADUCUS*, ready to fall) when a part falls off very early, compared with other parts with which it is associated. Thus the sepals of many Poppies fall as soon as the flower begins to expand.

CÆNO'BIO, synonyme for *CARCE'RULUS*.

CÆSPITO'SE, *CÆSPITO'SUS*, (*CÆSPES* turf) where plants are densely crowded in turf-like patches.

CÆTO'NIUM, synonyme for the Gluma of grasses.

CALA'MUS, (a reed) has been restricted to hollow inarticulate stems, like those of rushes.

CALATHI'DIUM, *CA'LATHIS* (καλαθις a basket) the peculiar form of the Capitulum, assumed by the Compositæ. Synonyme for Anthodium.

CALCAR, a spur.

CALCARA'TUS, (*CALCAR* a spur) furnished with a spur.

CALCA'REUS, (pertaining to lime) of a dull chalk-white colour.

CALCEIFOR'MIS, (*CALCEUS* a shoe, *FORMA* shape) where an organ is inflated and shaped somewhat like a shoe, as the lip of Cypripedium.

CALLITRICHI'NEÆ, (from the genus Callitriche) a natural order of Dicotyledones, containing only one genus.

CALY'BIO, (καλυϐιον a cottage) synonyme for Glans.

CALYCA'TUS, furnished with a calyx; or where the calyx is large or remarkable.

CALYCE'REÆ, (from the genus Calycera) a small natural order of Dicotyledones, closely allied to Compositæ.

CALYCIFLO'RÆ, an artificial group, formed from those orders of Dicotyledones, where the stamens adhere to the calyx; whether they are perigynous, or epigynous.

CALYCENA'LIS, belonging to the calyx; as the pubescence &c. upon it.

CALYCINIA'NUS, formed, or derived from the calyx; as the induvies of certain fruits. Ex. gr. the Rose.

CALYCI'NUS, of the nature or appearance of a calyx; as some involucres. Also used as a synonyme for *CALYCATUS*.

CALY'CULATE, *CALYCULA'TUS*. When the flower appears as though it possessed a double calyx, applied especially to the outermost bracts of certain Compositæ, which stand apart from the rest of the involucrum, and appear to form a distinct whorl of themselves, which is termed a calyculus.

CALYP'TRA, (καλύπτρα) a Veil.

CA'LYX, *CA'LYX* (καλυξ the cup of a flower) the outermost whorl of the perianth, composed of the sepals, either free or cohering. When the perianth consists of a single floral whorl, it is generally considered as a calyx, rather than a corolla; though formerly it frequently went by the latter name, when it happened to be highly coloured, (as in the Tulip,) or did not possess the more usual green and leafy appearance of the calyx.

CAM'ARA, (καμαρα, a vaulted chamber) a fruit where the pericarp is more or less membranous, and consists of two adhering valves, with one or more seeds attached to the inner angle, as in the Ranunculaceæ, the core of the apple, &c. This definition includes several very distinct forms of fruit.

CAM'BIUM, a highly viscous fluid, elaborated by the internal organs of plants, and serving for the nourishment of their several parts. The term is more especially applied to the clammy secretion, formed in spring, between the bark and wood of Dicotyledonous trees.

CAMPANULA'CEÆ, (from the genus Campanula) the Campanula Tribe. A natural order of monopetalous Dicotyledones.

CAMPAN'ULATE, *CAMPANULA'TUS.* (*CAMPANA*, a Bell) Bell-shaped, as the corolla of Campanula.

CAMPULI'TROPOUS, *CAMPULI'TROPUS*, (καμπυλος curved, τρεπω to turn) where the ovule and its integuments are so bent that the apex is brought near the hilum. The hilum and chalaze being together, fig. 43.

CANALICULA'TUS, channelled.

CANCEL'LATE, (*CANCELLA'TUS* latticed) where there is an appearance somewhat resembling lattice-work; as where the single fibres, of which the whole plant of Byssus cancellatus is composed, cross each other; or where the parenchyma in the leaves of Hydrogeton being wanting, the nerves only are left with open spaces between them.

CAN'DIDUS, (white) a pure white, but not so clear as snow-white.

CANES'CENS, *CANUS* (hoary) more or less grey, verging on white.

CAPILLA'CEOUS, *CAPILLA'CEUS* (hairy, or like hair) as fine as hair.

CAPILLAMENTO'SUS, comose.

CAPILLAMEN'TUM, synonyme for filamentum.

CAPILLA'RIS CAPILLA'TUS (of, or like hair) of the form, as well as about the size of hair.

CAPILLI'TIUM, filamentous tissue, among which, the sporules of certain fungi are disposed in their state of fructification.

CAP'ITATE, *CAPITA'TUS* (having a head) where the summit of some slender organ, as hair, the style, &c., is swollen out, or appears to be capped by an expansion, somewhat like a head on a pin, fig. 44.

CAPI'TULUM, (a little head) where the inflorescence consists of numerous flowers, sessile or nearly so, collected into a dense mass at the summit of the peduncle.

CAPREOLA'TUS, (*CAPRE'OLUS* a tendril of a vine) bearing tendrils.

CAPRE'OLUS, synonyme for Cirrhus; a Tendril.

CAP'SULE, *CAP'SULA* (a chest) a dry dehiscent seed vessel, with one or more cells, and many seeds; as in the Primrose and Rhododendron. This term has also been applied to the Anther; and still more commonly to the Theca of Ferns and Mosses.

CAP'SULAR, *CAPSULA'RIS*, related to a capsule; or having a capsule in someway remarkable.

CARCE'RULA-LUS, (*CARCER* a goal) a dry indehiscent many-celled fruit, with few seeds in each cell; the cells cohering round a common style, placed in the axis. Examples are Tilia, Malva.

CARI'NA, a keel.

CARINA'TED, *CARINA'TUS*, keeled.

CARIOP'SIDE, *CARIOP'SIS* (κάρη the head, οψις appearance) a dry one-seeded indehiscent fruit, in which the endocarp adheres to the spermoderm. The fruit of grasses.

CAR'NEUS, (*CARO*, *CARNIS* flesh) pale red, of a flesh colour.

CA'NOSUS, fleshy.

CA'RO, (flesh) firm pulp, eatable part of the melon.

CARPADEL'IUM, (καρπος, fruit and δηλος, hidden) synonyme for *CREMOCARP'UM*.

CAR'PEL, *CARPE'LLA* (καρπος, fruit) one of the subordinate parts, whether free or cohering, which compose the innermost of the four sets of floral whorls, into which the complete flower is separable. It bears the same relation to the gynæceum as a sepal to the calyx, and a petal to the corolla.

CARPO'PHORUS,-RUM, (καρπος fruit, and φερω to bear) synonyme for gynophorus.

CARTILA'GINOUS, *CARTILAGI'NEUS* (of a gristly substance) tough and hard, like the skin on the pip of an apple.

CARUN'CULA, (a little piece of flesh) a swollen fungous-like excrescence on the surface of some seeds, about the hilum.

CARYOPHILLA'CEOUS, *CARYOPHILLA'TUS* (like the petals of *DIANTHUS CARYOPHILLUS*) a flower whose five petals have long narrow claws.

CARYOPHYL'LEÆ, (from Caryophyllus, the specific name of the clove-pink) the Pink-Tribe. A natural group of Dicotyledones, which may be considered as a distinct order, or be subdivided into certain subordinate groups, each of which may be so considered.

Caryop'sis, see *Cario'psis*.

Cassid'eous, *Cassi'deus* (*Cassis*, a helmet) where a very irregular flower, as the Aconite, has one large helmet-shaped petal.

Cassytha'ceæ, Cassy'theæ, a natural group of Dicotyledones, containing only the single genus Cassytha; and which may be considered either a distinct order, or a tribe of Laurineæ.

Casuara'ceæ Casuari'neæ, a natural order for Dicotyledones, including only the genus Casuarina.

Catacle'sium, synonyme for Diclesium.

Catapet'alous, *Catapeta'lus* (κατὰ under, πεταλον, a petal) where the petals of a polypetalous corolla adhere to the bases of the stamens; as in Malva.

Catkin, a peculiar form of spiked inflorescence where the flowers are unisexual, closely crowded, and the place of each perianth is supplied merely by a bract, as in Salix, Corylus, &c. See fig. 45.

45

Cat'ulus, synonyme for the more frequently used term *Amentum*, a Catkin.

Cau'da, a tail.

Cauda'tus, tailed, or tail-pointed.

Caudex, (*Caudex*, the trunk) the main trunk of the root. Also, the stem of Palms, and Tree-ferns.

Caudi'cula, (*Cauda*, a tail) an elastic appendage to the pollen masses of certain Orchidaceæ, derived from the remains of the cellular tissue, in which the pollen grains are developed. See fig. 46.

46

Caules'cent, *Caule'scens* (*Caulis*, a stalk) where a stalk is distinctly visible.

Caulicu'le, *Cauli'culus* (a little stalk) synonyme for neck.

Cauli'culi, small stems rising immediately from the neck of the root.

Cauli'nar, *Caulina'ris*. *Caulina'rius*, *Cauli'nus*, (*Caulis*, a stalk) belonging to the stem, or growing from it.

Caulis, a stem.

Caulocar'pous, *Caulocar'peus*, *Caulocar'picus* (καυλὸς the stem, καρπὸς fruit) a term for trees and shrubs, whose woody stem and branches do not die away, as the herbaceous ones of perennials, but continue to bear flowers and fruit for a succession of years.

CEDRELA'CEÆ CEDRE'LEÆ, (from the genus Cedrela) a natural group of Dicotyledones, considered to be either a section of Meliaceæ, or a distinct order.

CELASTRA'CEÆ, CELASTRI'NEÆ (from the genus Celastrus) a natural order of Dicotyledones.

CELL, *CEL'LA*, *CEL'LULA* (a little cellar) each of the vesicles of which the cellular tissue is composed.

CELLULAR-TISSUE, an aggregation of minute membranous vesicles, of various sphæroidal or polygonal shapes, filled with fluid: and of which the main bulk of all vegetables is composed.

CENTRI'FUGAL, *CENTRI'FUGUS*, (*CENTRUM* a centre, *FUGO* to fly from) where an organ extends from the centre towards the circumference; as the radicle in the seed of Cucurbitaceæ. The inflorescence is so termed, when those flowers first expand which are seated nearest the main axis, and then those which are the next outermost in succession.

CENTRI'PETAL, *CENTRI'PETUS*, (*CENTRUM* a centre, *PETO* to seek) where an organ points from without towards the centre, as the radicle in the seed of Œnothera. The Inflorescence is termed centripetal when the flowers which are the lowest on the peduncle, or most distant from the main axis are evolved first, and then those which are next in succession.

CEPA'CEOUS, *CEPA'CEUS*, (*CEPA* an onion) a synonyme for alliaceous.

CEPHALAN'THIUM, (κεφαλὴ a head, ἄνθος a flower) synonyme for Anthodium.

CEPHALO'DIUM, (κεφαλὴ a head) synonyme for Tuberculum.

CEPHALOTA'CEÆ CEPHALO'TEÆ, a natural order of Dicotyledones, founded on the genus Cephalotus.

CERA'CEUS, *CERE'US*, waxy.

CERA'TIUM, (κέρας a horn) a long one-celled pericarp, with two valves, and containing many seeds attached to two placentæ, which are alternate with the lobes of the stigma. Ex. Glaucium.

CERATOPHYL'LEÆ, a natural order of Dycotyledones, including only the aquatic genus Ceratophyllum. It may possibly be considered to be a sub-order of Urticaceæ.

CEREA'LIS, (belonging to corn) any of the Gramineæ whose seeds serve for corn.

Ce'rinus, (of a wax colour) impure yellow slightly tinged with red.

Ce'rio, synonyme for Cariopsis.

Cer'nuous, *Cer'nuus*, (hanging down the head) where the summit inclines slightly from the perpendicular, as the flowers of a Narcissus with respect to their peduncles.

Cervi'nus, (*cervus* a stag) dark tawny, or deep yellow with much grey.

Cestra'ceæ, Cestri'næ (from the genus Cestrum) a natural order of Dicotyledones.

Chaff, small membranous scales, the degenerated state of bracts, seated on the receptacle of many Compositæ.

Chaf'fy, furnished with chaff.

Chailletia'ceæ, Chaille'tiæ, (from the genus Chailletia) a natural order of Dicotyledones.

Chala'ze, *Chala'za* (χάλαζα a spot on the skin looking like a hailstone) the disk-like appearance, or scar, formed at the spot where the inner integument of the ovule (the secundine or tegmen) is united to the outer (the primine or testa); and through which a vascular chord (the raphè [r]) passes to nourish the nucleus, which is also attached by its base to the same spot. fig. c

Chan'nelled, hollowed out, so as to resemble a gutter; as in the leaves of Tradescantia virginica.

Chara'ceæ, (from the Genus Chara) the Chara Tribe. A small natural order of Acotyledones, containing only the aquatic genera Chara and Nitella.

Cha'racter, any peculiarity in form or property, affording a mark of distinction or of resemblance, in different plants. All the characters of a species taken collectively form its "Natural character." Those which it possesses in common with other species of the same genus are collectively termed its "Generic character." An "Artificial character" is limited to a few of the characters belonging to one or more particular organs, as the flower or fruit, without considering any of the characters belonging to other organs.

Charta'ceous, *Charta'ceus*, (*Charta* paper) thin, flexible, and membranous, resembling paper or parchment. As the pericarp of Anagallis arvensis; many leaves, &c.

Chenopodia'ceæ, Chenopo'deæ, (from the Genus Chenopodium) the Goosefoot tribe. A natural order of Monochlamydeous Dicotyledones, containing a large proportion of humble weeds.

CHLORANTHA'CEÆ, CLORAN'THEÆ, (from the genus Chloranthus) a small natural order of Dicotyledones.

CHLORAN'THIA, (χλωρὸς green, ἄνθος a flower) a monstrous development of the floral organs, where they become more or less green, and partially assume the character of leaves.

CHLOROPHYL'LA, (χλωρος green, φυλλον a leaf) the colouring matter in plants, especially in their leaves.

CHLOROPHYL'LÆ, (χλωρος green, φυλλον a leaf) a name given to those parasitic phanerogamous plants which, like the Missletoe, are furnished with green or yellowish leaves.

CHLORO'SIS, (χλωρος green) a disease, where the green colour of the plant disappears or assumes a very faint tint.

CHO'RION, *CHO'RION*, (χωρίον the external fœtal membrane) a fluid pulp composing the nucleus of the ovule in its earliest stage; and which is gradually absorbed during the development of the embryo.

CHRO'MULE, *CHRO'MULA*, (χρῶμα colour) the colouring principle in all parts of plants, used synonymously with Chlorophylla.

CHRYSOBALANA'CEÆ, CHRYSOBALA'NEÆ, (from the genus Chrysobalanus) the Cocoa-plum Tribe. A natural group of Dicotyledones; which may either be considered a distinct order or a subordinate section of Rosaceæ.

CICA'TRICULE, (*CICATRI'CULA*, a little scar) synonyme for Cicatrix.

CICATRISA'TUS, marked by scars.

CI'CATRIX, (*CICA'TRIX* a scar) the impression left at the spot where an organ was articulated to some part of a plant, as the leaflets to the petiole, the leaf to the stem.

CICHORA'CEÆ, (from the genus Cichorium) a natural group of Dicotyledones, sometimes considered as a distinct order, but more usually as a separate tribe of the extensive order Compositæ.

CI'LIÆ, (*CI'LIUM*, an eyelash) somewhat stiffish hairs, which form a fringe on the margin of an organ, as those on the leaf of Sempervivum tectorum. Applied also to the teeth of the inner peristome of Mosses.

CILIA'RIS, *CILIA'TUS*, furnished with ciliæ.

CINCHONA'CEÆ, (from the genus Cinchona) the Coffee Tribe. A natural order of Dicotyledones; otherwise considered a tribe of the more extensive order Rubiaceæ.

CINE'REOUS, (*CINERE'US*, ash-colour) the intermediate tint

between pure white and black. *Cineraceus* implies it to be a little paler, and *Cinerascens*, very pale, bordering on white.

Cin'gulum, (a girdle) the neck of a plant.

Cinnabari'nus, (*Cinna'baris*, vermillion) scarlet slightly tinged with yellow.

Cinnamo'meus (*Cinnamo'mum* cinnamon) of a bright brown colour, formed from reddish-orange and grey.

Circina'lis, synonyme for *Circina'tus*.

Cir'cinate, *Circina'tus*, (*Cir'cino*, to turn round) Rolled inwards from the summit towards the base, like a Crozier. As the fronds of ferns in vernation. fig. 48.

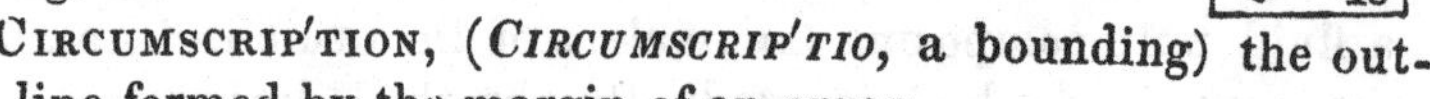

48

Circumscrip'tion, (*Circumscrip'tio*, a bounding) the outline formed by the margin of an organ.

Cir'rhous, (*Cirra'tus* curled) either furnished with a tendril (*Cir'ratus*), as the leaves of Gloriosa superba; or assuming the form and functions of a tendril (*Cirrho'sus*, *Cirro'sus*) as the peduncles of Clematis cirrhosa; or, where the tendrils are in some way remarkable (*Cirrha'lis*) as the ascidia cirrhalia of Nepenthes. These several terms are not in general sufficiently restricted, but have been used indiscriminately.

Cir'rhus, (*Cir'rhus*, a curl) synonyme for a Tendril.

Cista'ceæ, (from the genus Cistus) the rock-rose tribe. A natural Order of Dicotyledones.

Cistel'la, *Cis'tula*, (a little chest) when the Apothecium of Lichens, as in the genus Sphærophoron, is globular and closed in its early state (a), but bursts irregularly in maturity (b) and then discloses the mass of sporules arranged round a common centre. fig. 49.

49

Cisti'neæ, synonyme for Cistaceæ.

Cit'reus, *Cit'rinus*, (colour of Lemon) pure yellow very slightly tinged with grey.

Class, *Clas'sis*, one of the primary or largest groups under which plants are classified. In the Artificial System of Linneus these amount to 24. In the Natural System of Jussieu there are 3. In the Natural arrangement of Dr. Lindley there are 4.

Clav'ate, *Clava'tus*, *Clavella'tus*, *Clavifor'mis*, *Clavillo'sus*, (*Clava* a club) where any organ, slender at the base, gradually thickens towards the apex. As the filaments of Thalictrum clavatum, the stigma of Campanulæ.

CLAW, the narrowed base of some petals, analogous to the footstalk of leaves. Ex. Dianthus.

CLINAN'DRIUM, (κλὶνη a bed, ἄνηρ a man) a depression in the summit of the gynostemium of certain Orchidaceæ, either above or below the stigma, in which the anther is lodged.

CLINAN'THIUM, (κλὶνη a bed, ἀνθος a flower) a "receptacle" (of the flowers) which is not of a fleshy consistency. As in the Compositæ, fig. 50, *a*. N. B. The involucral scales in front are removed to show the Clinanthium.

CLOS'TER, *CLOS'TRUM*, (κλωστηρ a spindle) elongated cells, pointed at each end, and either cylindrical or fusiform, which enter largely into the composition of the woody parts of trees, &c.

CLOUD'ED, where a pale ground is partially obscured by ill-defined patches of a darker tint, gradually softening into it.

CLOVE, a name given by gardeners to the young bulbs which are developed about the old ones.

CLUB-SHAPED, synonyme for clavate.

CLUS'TER, synonyme for raceme,

CLUS'TERED, where numerous similar parts are collected in a close compact manner. As the flowers of Cuscuta.

CLYPEA'TUS, *CLYPEA'STRIFORMIS*, *CLYPEIFOR'MIS*, *CLYPEOLA'RIS*, (*CLY'PEUS*, a shield, *FORMA* a shape) scutate, or shield shaped.

COACERVA'TUS, (*COACER'VO*, to heap up) clustered.

COADNA'TUS, *COADUNA'TUS*, (same as *CONA'TUS*,) cohering.

COALES'CENS, *COA'LITUS*, (*COALES'CO* to grow together) cohering.

COA'RCTURE, *COARCTU'RA*, (*COARCTO*, to press together) synonyme for the Neck.

COATED, where the external parts are harder than the internal; or are composed of a distinct layer, as the bark on the trunk, the rind of fruit, &c.

COBŒA'CEÆ, (from the genus Cobœa) a group of dicotyledons, including the single genus Cobœa, considered as a natural order by some, but referred to Polemoniaceæ by others.

COBWEBBED, where the pubescence is composed of thin white and very long hairs, which are matted together over the surface like the web of the gossimer spider. Ex. Sempervivum arachnoideum.

Cocci'neus, (scarlet or crimson) red with a slight admixture of yellow.

Coc'cum, (a seed which was used for dying cloth of a scarlet colour.) The closed cells of a plurilocular pericarp which separate from each other when ripe, thus forming, as it were, so many distinct pericarps. By some, the term is restricted to such cells or carpels as these which further open with elasticity by an internal longitudinal suture. As in Euphorbia, Ricinus &c. Fig. 51.

51

Cochlear, (*Cochlea'ris*, belonging to a spoon) applied to an estivation where one part of the perianth is helmet shaped, larger than the rest which it entirely surrounds; as in Aconitum.

Coch'leate, *Cochlea'tus*, (*Coch'lea* a snail shell) spirally twisted like a snail shell. As the legumes of many species of Medicago, fig. 52.

52

Codio'phyllus, (κωδιον a fleece, φυλλον a leaf) where the leaf is covered with a woolly pubescence.

Cœlos'permum, (κοῖλος hollow, σπέρμα a seed) a seed in which the albumen is so curved that the base and apex approach each other. As in a group of Umbelliferæ, named Cœlospermeæ.

Cæto'nium, (κοιτων a bedchamber) a name which has been given to the outer glume of a multifloral spikelet in grasses.

Cohe'ring, *Cohæ'rens*, (*Cohæ'reo*, to stick together) where similar parts are fastened together, as where the subordinate parts or petals of a corolla unite, so as to form into a tube, commonly called a monopetalous corolla. The term is often used in a more lax manner, as synonimous with "accrete" and "adhering."

Co'hort, *Co'hors*, (a band of soldiers) a group of plants formed by uniting several Orders together: a subdivision of a Class containing one or more Orders.

Colchica'ceæ, (from the genus Colchicum) a synonyme for Melanthaceæ.

Coleophyl'lum, (κολεὸς a sheath, φυλλον a leaf) a membranous or fleshy sheath investing the plumule in monocotyledoncus plants.

Coleop'tila, *Coleo'tilum*, (κολεὸς a sheath, πτῖλον a feather) synonymes for coleophyllum.

Coleorhi'za, (κολεὸς a sheath, ῥίζα a root) the substance of the radicular extremity in monocotyledonous seeds, through which the radicle bursts, in germination.

Coll'are, (a collar) synonyme for ligula.

Collect'ors, *Collecto'res*, (*Collect'us* gathered together) Papillary hairs on the style of Compositæ, Campanulaceæ, &c. whose use appears to be, to collect the grains of pollen, as the style elongates and forces the stigma past the anthers.

Col'lum, (a neck) the plane between the stem and root, termed the neck of a plant.

Col'oured, *Colora'tus*, (painted) when any part of a plant is not green; and sometimes when a subordinate part is differently coloured from the rest, though that may not be green either.

Co'lum, synonyme for placenta.

Columel'la, *Columnel'la*, (a little pillar) a persistent central axis, round which the carpels of some fruits are arranged, as in Geranium. Also the central axis in the thecæ of Mosses round which the sporules are seated.

Columellia'ceæ, Columelli'eæ, (from the genus Columellia) a natural order of Dicotyledones, including only the genus from which it derives its name.

Colum'na, (a pillar) the solid body formed by the union of the filaments, in some plants, as in Stapelia; a synonyme for gynostemium.

Co'ma, (a head of hair; also branches and leaves of trees) the aggregation of branches forming the head of a tree. A tuft of bracteæ (as in Fritillaria imperialis) or of barren flowers (as in Hyacinthus comosus) forming the summit or crown to the inflorescence. Also tufts of hair on certain seeds.

Coma'tus, (having hair) furnished with coma. Also applied to roots which are furnished with very numerous capillary ramifications.

Combreta'ceæ, (from the genus Combretum) a natural Order of Dicotyledones.

Combsha'ped, synonyme for pectinate.

Commellina'ceæ, Commeli'neæ, (from the genus Commelina) a natural Order of Monocotyledones.

Commissu'ra, (a knuckle joint) the inner surface of each of the two parts (mericarps) into which the fruit of the Umbelliferæ is divisable—where they mutually press against each other, fig. 53. (c) Also a point where many parts are united together.

COM'MON, *COMMU'NIS*, used synonimously with "general," "primary," "principal;" and in contradistinction to "partial." Where some part bears the same relation to several others, which another part does to each of them respectively: as the main petiole of a compound leaf, each of whose leaflets may have partial petioles of their own. The involucrum at the base of a compound umbel. The involucrum of the Compositæ was originally considered, and is still sometimes called, a common calyx; but with greater correctness the clinanthium is really a common receptacle.

COMO'SUS, synonyme for *COMA'TUS*.

COMPAC'TUS, (joined together) where the subordinate parts are very closely agglomerated or pressed together; as the several flowers composing the catkin of a Willow.

COMPLANA'TUS (made even or smooth) synonyme for *COMPRESSUS*.

COMPLE'TE, *COMPLE'TUS*, where no essential part is wanting; as where a flower is furnished with both stamens and pistils, that is, not unisexual. Also where an organ receives its fullest development, as where the dissepiments of a pericarp stretch from the paries to the axis, and thus distinctly divide it into separate cells.

COMPLEXI'VUS, (*COMPLEX'US*, embracing) when a leaf in vernation is folded over another, both at the sides and apex.

COMPO'SITÆ, (*COMPO'SITUS*, compounded; the aggregation of florets in the capitulum being formerly called a "compound flower") The most extensive Natural Order of Dicotyledones, strictly coinciding with the artificial Class Syngenesia. It has been proposed to subdivide it into four Orders, viz. Mutisiaceæ, Cichoraceæ, Asteraceæ, and Cynaraceæ.

COMPOUND', *COMPO'SITUS*, (set together, compounded) where the aggregation of several similar parts assumes the character of a distinct whole; as the combination of leaflets forming a compound leaf; or the aggregation of florets called a "compound flower" in the Compositæ; or the collection of carpels in a compound fruit.

CON'CAVE, (*CON'CAVUS*, hollow) applied to any surface with a curvilinear depression, or hollow formed without angles.

CONCEP'TACLE, *CONCEPTA'CULUM*, (a receptacle) synonyme for the thecæ or capsules of Ferns. Also a "double follicle" or fruit, composed of two follicular carpels which continue adhering (connate) as in Nerium, fig. (*a*) or soon separate (free), and appear as two follicles united at the base, as in Cynanchum, fig. (*b*) The side of one carpel is partially removed to shew the position of the seeds which detach themselves from the placenta and lie free; Fig. 54.

CONDU'PLICATE, *CONDUPLICATI'VUS*, (*CONDUPLICA'TUS*, doubled) in vernation, when the leaf is folded longitudinally, so that the parts of the surface on each side of the mid-nerve are parallel and close together. When applied to the cotyledons in the seed, it means that both having their upper surfaces in contact, are longitudinally folded together, and consequently in contrary directions, as in the radish.

CO'NE, *CO'NUS*, (a cone) a dense aggregation of scale-like carpels, arranged symmetrically round an axis, as in the Pine tribe.

COMPRES'SED, *COMPRES'SUS*, so flattened that two opposite surfaces are brought closer together than the others. As in the calyx of Rhinanthus, the pod of a Pea, or the stem of Poa compressa, where a transverse section presents an elliptic instead of a circular form.

CONFER'TUS, (full) crowded.

CO'NFLUENS, (*CONFLUO* to run together) cohering.

CONFOR'MIS, (*CONFOR'MO*, to make like to) where one part closely resembles another with which it is associated or compared.

CONGES'TUS, (heaped together) in vernation, where the leaves are folded up without regularity. Where the inflorescence is composed of flowers aggregated into a spherical head.

CONGLOMERA'TUS, (heaped together) clustered.

CO'NICAL, *CO'NICUS*, (*CO'NUS*, a cone) approaching the form of a true cone. A solid figure rising above a circular base into a point; as some prickles on the stems of roses, &c.

CONI'DIUM, (κονις dust) powdery particles which are aggregated in patches (*SOREDIA*) over the surface of the thallus of some lichens.

CONI'FERÆ, (*Co'nus* a cone) the Fir tribe. A natural order usually considered as part of the class Dicotyledones, the structure of the stem being exogenous. The cotyledons however, are very variable in number, and the structure of the seed-vessel, a carpellary scale, places them in the group Gymnospermæ, which is considered by some as a distinct Class.

CONIOCYS'TA, CONIOCYS'TIS, (χονία dust, χύστις a bladder) Apothecia which resemble a tubercle, and are filled with a mass of sporules.

CONIOTHE'CA, (κονία dust, θηκη a box) synonyme for the cell of an anther.

CON'JUGATE, *CONJUGA'TUS*, (*Conjugo* to couple together) a pinnate leaf, composed of a single pair of leaflets.

CONNARA'CEÆ, (from the genus Connarus) a small natural order of Dicotyledones.

CONNA'TE, *CONNA'TUS*, (*Con* together, *Nascor* to grow) where the bases of two opposite leaves are united round the stem, so that this appears to pass through them.

55

CONNEC'TIVE, *CONNECTI'VUM*, (*Con* together, *Necto* to bind) a portion of the stamen, distinct from the filament, which connects the cells of the anthers together.

CONNI'VENS, (winking) converging.

CONOI'DAL, *CONOI'DEUS*, (κωνοειδης formed like a cone) approaching a conical form.

CONSER'VATIVE-ORGANS, the parts or organs of a plant employed in carrying on the function of nutrition; as the root, stem, and leaves.

CONTI'GUOUS, (*Conti'guus* adjoining) when two neighbouring parts are in contact through the whole length of their edges or surfaces; as the sepals of Raphanus, and the cotyledons of many species.

CONTI'NUOUS, (*Conti'nuus* without intermission) where there is no break or deviation from uniformity, in some peculiar arrangement of subordinate parts. The term is used in contradistinction to "interrupted."

CONTORT'ED, (*Contort'us*, entangled, wreathed) where a part is folded or twisted back upon itself, as the root of the Polygonum bistorta. In estivation, this term is applied when the subordinate parts of the corolla are set obliquely and overlap each other in succession, as in the Order Apocynaceæ.

CONTRACT'ED, (*CONTRACT'US*, narrow, shortened) either, where some part appears to be unusually narrow, as the throat of Verbena officinalis with respect to the tube; or, where the longitudinal development of some parts is so shortened that the whole seems crowded, as the compact panicle of Dianthus barbatus.

CON'TRARY, (*CONTRA'RIUS* athwart) where some part ranges in a directly opposite direction to some other with which it is compared; as the dissepiment with respect to the valves in a loculicidal dehiscence, fig. 56. Otherwise used as synonymous with "opposite."

56

CO'NUS, a cone.

CONVERGINER'VIS, *CONVERGINER'VIUS*, *CONVERGENTI-NERVO'SUS*, (*CON* together, *VERGENS* bending, *NERVUS* a nerve) where the primary nerves of a leaf, meeting at the base and apex, curve in a regular manner between these points, as in the Convallaria majalis This term is sometimes restricted to that modification of curvinerved leaves where there are no secondary nerves.

CONVER'GING (*CON* together, *VERGENS* bending) where certain parts, separate at their bases, gradually approach each other at their apices.

CON'VOLUTE, *CONVOLUTI'VUS*, (*CONVOLU'TUS* wrapped together) rolled up in a longitudinal direction, so that one edge overlaps the other, as the spathe of an Arum. Also in estivation, where one part is completely rolled round another.

CONVOLVULA'CEÆ, (from the genus Convolvulus) The Bindweed-tribe. A natural Order of Dicotyledones.

CORACI'NUS, (of a raven black) deep shining black.

COR'CULUM, (a little heart) synonyme for Embryo.

CORDATE, *CORDA'TUS*, *CORDIFOR'MIS*, (*COR* the heart) shaped like the figure of a heart on cards; the point of attachment being at the broader end, as in many leaves, fig. 57.

57

CORDIA'CEÆ, (from the genus Cordia) a small natural Order of Dicotyledones.

CORIA'CEOUS, *CORIA'CEUS*, (*CORIUM* leather) leathery.

CORIARIA'CEÆ, CORIARI'EÆ, (from the genus Coriaria) a small natural Order of Dicotyledones.

CORKY, resembling cork in texture.

CORMUS, (κορμος a stem) the swollen succulent bulb-like

mass which composes the stem of certain monocotyledones, as in the Crocus, &c., and which has been frequently termed a solid-bulb. It is a variety of the rhizoma or under ground stem.

CORNA'CEÆ, COR'NEÆ, (from the genus Cornus) The Dogwood Tribe. A natural Order of Dicotyledones.

COR'NEOUS, (*CORNEUS* horny) resembling horn in consistency, translucency, and elacticity; as the albumen of the date and many other seeds, where these properties are seen upon cutting off a thin slice.

CORNICULA'TUS, (*CORNI'CULA* a little horn) horned.

CORNU, a horn.

CORNU'TUS, synonyme for *CORNICULA'TUS*.

COROL'LA, (*COROLLA* a little crown) the floral whorl next in succession within the calyx. It is composed of subordinate parts termed petals, which are either free, or more or less united together into a tube, in which case the corolla is termed monopetalous. It is generally more highly coloured than the calyx, but in many plants it is entirely wanting, and then the calyx frequently assumes the more usual aspect of the corolla.

COROLLA'CEUS, petaloid.

COROLLA'RIS, formed of, or belonging to the corolla.

COROLLA'TUS, furnished with a corolla.

COROLLIF'ERUS, (*COROLLA* a corolla, *FERO* to bear) supporting the corolla, as the gynophorus of Dianthus.

COROLLIFLO'RÆ, (*COROLLA* a corolla, *FLOR* a flower) a group composed of those Dicotyledonous Orders where the corolla is monopetalous and hypogynous.

COROLLI'NUS, either seated on the corolla, as pubescence in Menyanthus; or resembling a corolla in structure and otherwise termed petaloid.

COROL'LULA, (diminutive of corolla) the corolla of a small flower or "floret."

CORO'NA, (a crown) an aggregation of appendages, free or united, seated upon the inner surface of the perianth. As the tubular appendage, otherwise termed a Nectary, in the genus Narcissus; or the scales at the bases of the limbs of the petals of Silene. The term has also been used synonymously with the "Eye" of Apples, Pears, &c., formed of the withered persistent limbs of the calyx; for the "Ray" of the capitulum in some Compositæ, and for the "ligula" of Gramineæ; further also as a synonyme for *CUCULLUS*.

CORO'NANS, *CORONA'TUS*, crowning

CORO'NA STAMI'NEA, synonyme for *ORBI'CULUS*.

CORRUGATI'VA, *CORRUGA'TUS*, wrinkled.

COR'TEX, (rind or bark) used also for that part of the parenchyma of the leaf, which composes the outermost portion of the mesophyllum.

COR'TICAL, *CORTICA'LIS*, (*CORTEX* bark) adhering or belonging to the bark.

CORTICA'TUS, (furnished with a rind) coated.

CORTI'NA, (a curtain) a filamentous fringe round the margin of the pileus in Agarics, formed from the adhering debris of the veil.

COR'YMB, *CORYM'BUS*, (κόρυμβος the top) where the pedicels in the inflorescence, originate at different parts along the main axis, and elevate all the flowers to about the same height; the inferior pedicels being consequently longer than the upper ones, fig. 58.

58

CORYMBI'FERÆ, (*CORYM'BUS*, a corymb, *FERO* to bear) one of the three great groups considered by Jussieu as distinct Orders, into which he divided the extensive Natural Order Compositæ.

CORYMBO'SE, *CORYMBO'SUS*, approaching the form assumed by the corymb; and applied equally to the arrangement of the branches of some plants, and to the inflorescence of others.

COS'TA, (a rib) the midrib of the leaf.

COSTA'TUS, ribbed.

COTTONY, when the pubescence is composed of long, soft, hairs, which are entangled and interlaced, resembling raw cotton in appearance.

COTYL'EDON, *COTYLE'DON*, (κοτυληδὼν a hollow vessel) a part of the embryo, representing a first leaf, in the modified form in which it appears in the seed. Some embryos possess only one (Monocotyledones) others have two (Dicotyledones,) or even more Cotyledons.

COTYLEDONA'RIS, formed by the union or close approximation of the Cotyledons.

COTYLEDO'NEUS, possessing Cotyledons

COTYL'IFORM, *COTYLIFOR'MIS*, (κοτύλη a cavity, like a porringer, *FORMA* shape) synonyme for *COTYLOIDEUS*.

59

Cotyloi'deus, (κοτυλη & ειδος a resemblance) shaped something like a porringer, depressed in the middle, elevated on the margin and with a short wide tube at the base, fig. 59.

Crassula'ceæ, (from the genus Crassula) the House-leek tribe. A natural Order of Dicotyledones.

Cras'sus, thick.

Crate'riform, *Craterifor'mis*, (*crater* a goblet, *forma* shape) goblet-shaped.

Cream-colour, synonyme for Ivory-white.

Creeping-stem, an underground stem, which in its most general signification is synonymous with Rhizoma.

Cremoca'rp, *Cremoca'rpium*, (κρεμαω to hang, καρπὸς fruit) the fruit of Umbelliferæ, consisting of two one-seeded carpels, completely invested by the tube of the calyx, which forms an outer skin. When ripe, the carpels (called mericarps) separate, and are then suspended from the summit of a central slender branched column which was previously concealed. See fig. 53. The term has been extended to other fruits with more than two cells but of nearly similar structure in other respects.

60

Crena'te, *Crena'tus*, (*crena* a knotch) any surface or edge which presents a series of rounded prominencies; but more especially the margin of a leaf jagged in a regular manner with rounded teeth, fig. 60.

Cren'el, (*Cre'na* a notch) a rounded tooth in the crenate margin.

Cren'elled, *Crenula'ris*, *Crenula'tus*, synonyme for crenate.

Cres'cent-shaped, approaching the figure of a Crescent; as the glands on the involucrum of Euphorbias. Certain leaves, &c.

Crest'ed, surmounted by an irregular crest-like appendage.

Creta'ceous, (chalky) chalk-white.

Cri'nitus, (*Crinis* hair) bearded.

Crispati'vus, *Crispa'tus*, *Cris'pus*, curled.

Crista'tus, crested.

Cro'ceus, saffron-coloured.

Crook'ed, synonyme for curved.

Crowded, when subordinate parts thickly surround a common support or axis.

CROWN'ING, when prominently placed on the summit or apex of any part.

CRU'CIATE, *CRUCIA'TUS*. (*CRUX* a cross) when parts, set in opposite pairs, are so arranged round an axis that the consecutive pairs are at right angles to each other. As the leaves of *GALIUM CRUCIATUM*. The arrangement is similar to that in fig. 40, under the term Brachiate. Also a synonyme for Cruciform.

CRUCIFE'RÆ, (*CRUX* a cross, *FERO* to bear) A natural order of Dicotyledones: thus named from the four petals being placed in opposite pairs, with their limbs expanded so as to form a cross.

CRU'CIFORM, *CRUCIFOR'MIS*, (*CRUX* a cross, *FORMA* shape) where any parts in the same horizontal plane are disposed in the form of a cross; as the petals of Cruciferæ.

CRUEN'TUS, CRUENTA'TUS, (made bloody) marked with red blotches; also where any part is wholly red.

CRUS'TA, (κρυος cold) a granular, frosted thallus, peculiar to some Lichens; with a resemblance to hoar frost.

CRUSTA'CEOUS, *CRUSTA'CEUS*, *CRUSTA'TUS*, (*CRUSTA* a crust) hard and brittle, as the covering to the seed of Ricinus: also, "resembling a crust," as certain cryptogamic plants which encrust the surface of other bodies

CRYP'TA, (a vault) receptacles for the oily and other secretions of plants; like those which occur in the leaves of the Myrtaceæ.

CRYPTOGA'MIA, (κρυπτω to hide, γαμος a marriage) a Class in the artificial system of Linneus, including all the lower tribes of plants which are not furnished with true flowers; corresponding to the natural order Acotyledones.

CRYPTOGA'MIC, *CRYPTOGA'MICUS*, *CRYPTO'GAMUS*, with the characteristic of the Cryptogamia, viz. no true flowers, or at least so indistinct as to be very different from the flowers of Cotyledonous plants.

CRYPTOPHY'TUM, (κρυπτω to hide, φυτον a plant) a term which has been applied to some of the lowest tribes of cryptogamic plants—of which the organization is least understood.

CU'BICAL, *CU'BICUS*, approximating to a cubic form.

CU'BIT, (*CU'BITUS*, a measure, about half a yard in length) roughly estimated at about the length between the elbow and the tips of the fingers.

CU'BITAL, *CUBITA'LIS*, the length of a cubit.

CU'CULLATE, *CUCULLA'RIS*, *CUCULLA'TUS*, *CUCULLIFOR'MIS*, (*CUCUL'LUS*, a hood, *FORMA* shape) where a plane surface, as of a leaf, petal, &c. is rolled up like a cornet of paper; as the spathe of an Arum; fig. 61. Synonyme for hooded.

CUCUL'LUS, a hood.

CUCURBITA'CEÆ, (from the genus Cucurbita) the Gourd or Cucumber Tribe. A natural Order of Dicotyledones.

CULM, *CUL'MUS*, (corn straw) synonyme for straw.

CUL'TRATE, *CULTRA'TUS*, *CULTRIFOR'MIS*, (*CULTER* a knife, *FORMA* shape) approaching the shape of a knife-blade.

CU'NEATE, *CUNEA'RIUS*, *CUNEA'TUS*, *CUNEIFOR'MIS*, (*CUNEUS* a wedge, *FORMA* shape) wedge-shaped.

CUNONIA'CEÆ, (from the genus Cunonia) a natural Order of Dicotyledones.

CUPOLA-SHAPED, nearly hemispherical, like the cup of an acorn; fig. 62.

CU'PREUS, (made of copper) of a copper colour, yellowish-red with a considerable mixture of grey.

CUP, the receptacle of the fructification in certain Lichens; synonyme for "shield."

CUP-SHAPED, concave, hemispherical, and tapering below like a drinking cup.

CU'PULA, (a little cup) an involucrum composed of bracts which adhere together by their bases, and form a sort of cup in which the fruit is seated; as in the Oak, Beech, Nut, &c.

CUPUL'ARIS, (*CU'PULA* a little cup) formed like a cup.

CUPULA'TUS, furnished with a *CU'PULA*.

CUPULI'FERÆ, (*CUPULA*, and *FERO* to bear) the Nut Tribe. A natural Order of Dicotyledones.

CUPULIFOR'MIS (*CUPULA* a little cup, and *FORMA* shape) Cupola-shaped.

CURL'ED, when a foliaceous organ is irregularly folded and crimped; as the leaves of the cultivated variety of endive.

CUR'VATIVE, *CURVATI'VUS*, (*CURVO* to bend) in vernation and estivation, where the separate parts are scarcely folded but have the margins merely curved a little.

CURV'ED, *CURVA'TUS*, (bent) bent in the form of a bow or arc of a circle; so that the extremities approach each other.

CURVE-RIBBED, synonyme for curvinerved.

Curve-veined, *Curvive'nius*, (*Curvus* a curve, *ve'na* a vein) synonyme for curvinerved.

Curviner'ved, *Curviner'vius*, (*Curvus*, a curve, *nervus* a nerve) more strictly applied to those leaves only, where several nerves, having nearly the same thickness, and diverging from the base, meet again by converging to the apex, as in theLilley of the Valley (Convallaria majalis,) fig. 63. But the term is extended, to include also those leaves where the nerves, collected in a thick bundle, form a strong mid-rib, with some of them diverging from it at intervals,and running in parallel and somewhat curved lines, to the margin, fig. 64, as in the Banana and Plantain. (Musa sapientum) The former description of curvinerved leaves are said to have their nerves "convergent," and the latter "divergent."

Cuscuta'ceæ, Cuscu'teæ, Cuscuti'næ, (from the genus Cus'cuta,) a natural group of Dicotyledones, including only the genus Cuscuta, which is by some considered to form a separate Order, and by others a section of Convolvulaceæ.

Cush'ion, a swollen part of the stem or branches immediately below the leaves, more particularly observable in the Leguminosæ.

Cush'ioned, more or less hemi-spheroidal and flattened above, resembling a cushion.

Cus'pidate, (*Cuspida'tus* pointed) gradually tapering into a sharp stiff point.

Cut, where the incisions are rather deep and regular; as those in the margins of leaves, which extend to a greater depth than where they are said to be "toothed," but not so deep as in "laciniate."

Cu'ticle, (*Cu'ticula* the outer skin) synonymous with epidermis. By some, however, restricted to the integument of the grains of pollen; by others to its external pellicle.

Cut'ting, a portion of a young branch which when inserted into the earth, and under suitable treatment, emits roots, and is developed as a distinct individual.

Cya'neus, (bright blue) pure blue.

Cyathifor'mis, (*Cya'thus* a cup, *forma* shape) cup-shaped

Cycada'ceæ, Cyca'deæ, (from the genus Cycas) a small natural order of Gymnospermæ.

CYCLANTHA'CEÆ, CYCLAN'THEÆ, (from the genus Cyclanthus) a natural order of Monocotyledones.

CY'CLICAL, (*CY'CLICUS* circular) completely coiled into a circle; as the embryo of Basella rubra.

CYCLOSIS, (κυκλος a circle) the partial circulation observable in the milky juices of certain plants—as in some of the genera Ficus, Chelidonium, &c.

CYLIN'DRICAL, *CYLIN'DRICUS*, *CYLINDRA'CEUS*, approaching closely to the form of a cylinder; as the stems of grasses, &c. which, however, all taper more or less, although by insensible degrees.

CYMA'TIUM, (κυμάτιον a little wave) synonyme for Apothecium.

CYMBÆFOR'MIS, *CYMBIFOR'MIS*, (*CYMBA* a boat, *FORMA* shape) boat-shaped.

CYME, (*CYMA* a sprout) an inflorescence, where numerous peduncles are given off in all directions from the summit of a branch, and the whole assumes the general appearance of a compound Umbel; but the peduncles are branched at different altitudes, and the pedicels are consequently of different lengths, since the flowers all stand at the same general level. Examples occur in the common Elder (*SAMBUCUS NIGRA*) *VIBURNUM*, &c. fig. 65. The term is restricted by some to an inflorescence whose main axis is terminated by a flower, and below this are two or more verticillate bracts, from the axils of which proceed lateral peduncles , on each of which a flower, fresh bracts and peduncles are then arranged similarly to those on the main axis, fig. 66. This latter mode of in florescence is otherwise termed a "*FASCICLE*." Th term has also been applied to the aggregate of the branches of a tree above the trunk.

CYMO'SUS, furnished with a cyme.

CYNARRHO'DUM, *CYNNARO'DIUM*, a fruit composed of several free, hard, and indehiscent ovaries, enveloped by, but not united to, the fleshy tube of the calyx. As in Roses.

CYNOMORIA'CEÆ, *CYNOMORI'EÆ*, (from the genus Cynomorium) a natural order of parasitical phenogamous plants, in the small class of Rhizantheæ.

CYPER'ACEÆ, CYPERO'IDEÆ, (from the genus Cyperus) the Sedge tribe. An extensive Order of Monocotyledones, with glumaceous flowers; whose general appearance approaches that of Grasses.

CYPHEL'LA, (κῦφος bent in, gibbous) orbicular fringed spots like dimples, confined to the under surface of certain lichens; as in the genus Sticta.

CYP'SELA, *CYP'SELLA*, (κυφελὶς a casket) a synonyme of Achenium; where that term is applied to the fruit resulting from an inferior ovarium, as in the Compositæ.

CYRTANDRA'CEÆ, (from the genus Cyrtandra) a natural order of Dicotyledones.

CYSTID'IUM, (κύστις a bladder) synonyme for the fruit termed a Utriculus.

CYS'TULA, a synonyme for cistella.

CYTINA'CEÆ, CYTI'NEÆ, (from the genus Cytinus) a natural order of parasitical phænogamous plants, of the class Rhizantheæ.

DACRYOI'DEUS, (δάκρυ a tear, εἶδος a resemblance) applied to a pear-like fruit, oblong and rounded at one end, and pointed at the other.

DACTYLO'SUS, (δακτυλος a finger) oblong and nearly cylindrical; as the spikes of Panicum dactylon.

DÆ'DALOUS, *DÆDA'LEUS*, (*DÆ'DALUS* artificial,) where the broad apex of a leaf, without being strictly truncate, is irregularly jagged.

DASYPHYL'LUS, (δασυς thick, hairy, φυλλον a leaf) where the leaves are either densely aggregated, or else covered with close woolly hair.

DATISCA'CEÆ, DATIS'CEÆ, (from the genus Datisca) a small natural order of Dicotyledones.

DEALBA'TUS, whitened.

DECAGY'NIA, (δεκα ten, γυνὴ a woman) an artificial order in the Linnean system, consisting of plants which have either ten pistils, or whose pistil has ten free styles.

DECAN'DRIA, (δεκα ten, ἀνὴρ a man) the tenth artificial class of the Linnean System, including certain flowers with ten stamens, not belonging to other classes. Decandrous flowers in the classes Monadelphia, Diadelphia, Monœcia, Diœcia, form the orders "Decandria," severally subordinate to those classes.

DECI'DUOUS, (*DECI'DUUS* subject to falling) applied to the leaf, it designates those trees and shrubs which shed all their leaves annually at the same period, so that their branches become quite bare; whereas in evergreens, the old leaves do not fall until the young ones are fully expanded. Applied to the calyx, corolla, or other organs, is signifies that these fall off sooner or later, after their functions have been performed, whilst other parts still remain, or are persistent.

DE'CLINATE, DECLI'NING, (*DECLINATUS* turned aside) where an organ or set of organs is bent or inclines towards one side; as the stamens of an Amaryllis with respect to the axis of the corolla. Also used synonymously with "deflexed, and "inclining."

DECOM'POUND, DE'COMPOUNDED, *DECOMPO'SITUS*, (*DE* from, *COMPOSITUS* compounded) where the principle of subdivision is carried to a considerable extent. Thus the stems of low shrubs, which are subdivided from the very ground into numerous branches, are said to be decompounded. Where secondary, tertiary, &c. petioles are formed in a leaf, so that it is more than simply compound, this organ is also said to be decompounded.

DECREAS'INGLY-PINNATE, *DECRESCEN'TE-PINNA'TUS*, where the leaflets of a pinnate leaf gradually decrease in size from the base towards the apex.

DECUM'BENT, (*DECUMBENS* lying down) applied to stems, when they recline upon the surface of the earth, but have a tendency to rise again towards their extremities. Applied to stamens, it is a synonyme for Declinate.

DECUR'RENT, (*DECURRENS* running down) where the limb of a leaf is prolonged down the stem on each side, below the point where the midrib quits it; as though the leaf were partially united to the stem by its midrib. Common in thistles, Carduus. Fig. 67.

DECUR'SIVELY-PINNATE, *DECURSIVE-PINNATUS*. (*DECURSUS* a descent) where a leaf has a decided appearance of being pinnate, but the leaflets are decurrent along the petiole.

DECURSI'VUS, decurrent.

DECUS'SATE, *DECUSSATI'VUS*, (*DECUSSATUS* cut cross-ways) synonyme for brachiate (fig. 40); but more especially applied to the leaves.

DE'FINITE, (*DEFINITUS*, defined, finite) applied to the stamens when they do not exceed twelve in number, and are constant in the same species. Used also synonymously with "terminal," for that particular kind of inflorescence

Defix'us, (fastened) synonyme for "Immersus."

Deflex'ed, (*Deflexus* bended) bending gradually downwards through the whole length.

Deflora'tus. (*De* from, *floreo* to blossom) the condition of the anther after having discharged its pollen.

Degenera'tion, *Degenera'tio*, (*Degenero* to degenerate) some peculiarity in the condition of an organ, induced by a modification of the circumstances under which its more usual and healthy development is effected.

Dehis'cence, *Dehiscen'tia*, (*Dehisco* to gape) the manner in which an organ, closed at first, ultimately bursts; but more especially applied where the bursting is with regularity along particular lines of suture; as in the anthers for the discharge of the pollen; in many pericarps for the escape of the seeds.

Dehis'cent, *Dehis'cens*, bursting by a regular dehiscence.

Deliques'cent, (*Deliquescens* melting down) where a main axis is lost in numerous subdivisions; as in the repeated branching of many stems; in the ramification of peduncles into numerous pedicels.

Del'toid, *Deltoi'deus*, (δελτα the Greek letter Δ and ειδος a resemblance) applied to succulen leaves, whose transverse sections have a resemblance to a Δ. See figure 68.

68

Demer'sus, (drowned) applies to those parts of an aquatic which are constantly below the surface of the water.

Demis'sus, (hanging down) lowered.

Dendri'ticus, *Dendroi'des*, (δενδρον a tree, ειδος a resemblance) assuming the general form of a tree. Also, having somewhat of a shrubby character.

Dens, a tooth.

Den'tate, (*Dentatus* toothed) synonyme for toothed.

Den'ticulatus, furnished with small teeth.

Denuda'tus, made naked.

Depaupera'tus, (impoverished) starved.

Depen'dens, (hanging down) pendent.

Depres'sed, (*Depressus* pressed down) where the longitudinal extension is much smaller than the transverse. Having the appearance of being flattened vertically, as the tuber of the turnip.

Descend'ing, *Descen'dens*, tending gradually downwards; as some branches and leaves. Also, penetrating more or less vertically into the earth; as with the root, the descending axis of vegetation.

DESVAUXIA'CEÆ, DESVAUX'IEÆ, (from the genus Desvauxia) a small natural order of Monocotyledones.

DETEC'TUS, naked.

DETER'MINATE, (*DETERMINATUS* limited) synonyme for Definite.

DEVEL'OPMENT, (*DEVELO* to open, unveil) that gradual extension of parts by which any organ or plant proceeds from its nascent state to maturity.

DEVIA'TUS, (*DE* from, *VIA* a way) reversed, as where the upper surface of a leaf is turned towards the ground, and the under towards the sky.

DEW'Y, where a surface appears as if covered with dew, arising from small irregular and pellucid expansions of cellular tissue.

DEXTROR'SUM, (towards the right hand) applied to a spiral whose successive convolutions would appear to a person, placed in its axis, to rise from left to right; as in the Hop, (*Humulus lupulus.*) See figure 69.

69

DI'ACHYMA, (δια through, Χυμιος juice) the parenchyma of leaves, occupying the space between their two surfaces.

DIADEL'PHIA, (δὶς two, αδελφος a brother) an artificial class in the Linnean system, characterized by the stamens being united by their filaments into two distinct bundles.

DIADEL'PHOUS, *DIADEL'PHICUS*, *DIADEL'PHUS*, where the stamens are united into two distinct bundles, as in Diadelphia.

DIANDRIA, (δις two, ανηρ a man) an artificial class in the Linnean system, containing plants whose flowers have only two stamens.

DIAN'DROUS, *DIAN'DER*, *DIAN'DRUS*, a flower containing only two stamens.

DIAPENSIA'CEÆ, (from the genus Diapensia) a small natural order of Dicotyledones.

DICHO'TOMOUS, *DICHO'TOMUS*, (διχοτομος divided into two) where any part forks or subdivides into two branches, and each of these again into two others; as the stems of Stellaria holostea; the leaves of Ceratophyllum demersum. Fig. 70.

70

DICLE'SIUM, *DYCLE'SIUM*, (δις two, κλειω to shut) a fruit composed of an indehiscent one-seeded pericarp, invested by a persistent and indurated perianth. Ex. Mirabilis.

DICLI'NIS, (δὶς two, κλίνη a bed) synonyme for Dioicus.

DICOTYLEDO'NES, (δὶς two, κοτυλήδων a seed-leaf) a natural class, containing plants whose seeds have two cotyledons, or more.

DICOTYLE'DONOUS, *DICOTYLEDO'NEUS*, possessing two cotyledons.

DI'DYMUS, (διδυμος double) twin.

DIDYNA'MIA, (δις two, δυναμος power) an artificial class in the Linnean system, characterized by the flowers being irregular, and containing four stamens, of which two are longer than the other two. Fig. 71.

DIDYNA'MOUS, *DIDYNA'MUS*, where the stamens are arranged as in Didynamia.

DIERE'SILIS, (διαίρεσις a division) synonyme for Carcerula.

DIFFOR'MIS, (*DE* from, *FORMA* a form) having an unusual shape. Or, remarkable for some singularity of shape.

DIFFU'SE, (*DIFFU'SUS* spread abroad) spreading widely, horizontally, and irregularly.

DI'GAMUS, (δις twice, γαμος a marriage) when two kinds of flowers, some male and the others female, are placed on the same receptacle, in Compositæ.

DIGITA'LIFORMIS, fox-glove shaped.

DIGITA'TE, (*DIGITA'TUS* having fingers) (1.) applied to a simple leaf; where the lobes are very narrow, deeply cut, and all extend nearly to the base of the limb, imitating the fingers of the human hand, (Fig. 72. a). (2.) In compound leaves, where the leaflets are all placed at the very extremity of the petiole, (b.)

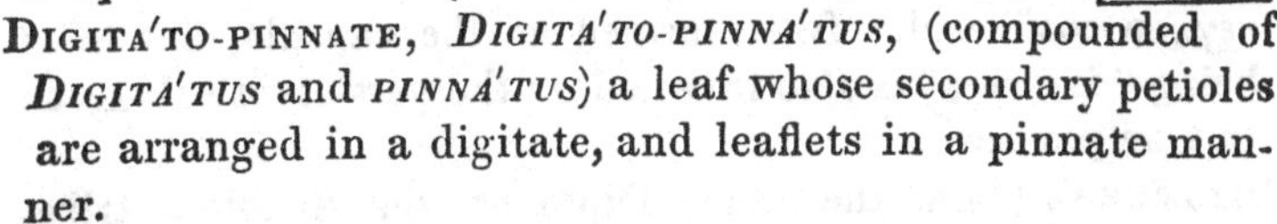

DIGITA'TO-PINNATE, *DIGITA'TO-PINNA'TUS*, (compounded of *DIGITA'TUS* and *PINNA'TUS*) a leaf whose secondary petioles are arranged in a digitate, and leaflets in a pinnate manner.

DIGITINER'VIUS, (*DI'GITUS* a finger, *NERVA* a nerve) where all the secondary nerves or ribs of a leaf diverge from the summit of the main petiole; as in fig. 72, b. Straight-ribbed.

DI'GITUS, (a finger) expresses about three inches in length.

DIGYN'IA, (δις twice, γυνη a woman) an order, in some of the classes of the artificial system of Linneus, characterised by the flowers having two pistils, or at least two distinct styles.

DIGY'NOUS, *DIGY'NUS*, (δις twice, γυνη a woman) either possessing two distinct pistils; or a pistil with two distinct styles; or, with two distinct stigmas.

DILA'TED, (*DILATA'TUS*) expanding into a lamina, as if flattened out, as the filaments of an Ornithogalum.

DILLENIA'CEÆ, (from the genus Dillenia) a natural order of Dicotyledones,

DILU'TUS, (washy, thin) any colour of a very pale tint.

DIMID'IATE, (*DIMIDIA'TUS* halved) where partial imperfection seems to exist; as in a stamen whose anther has only one lobe; a leaf whose limb is fully developed on one side of the mid-rib, and scarcely at all on the other. Also applied to the veil (calyptra) of mosses, when it is longitudinally split on one side, by the swelling of the theca.

DIMOR'PHOUS, *DIMOR'PHUS*, (δις twice, μορφη form) where similar parts of the same plant assume different shapes, or characters.

DIŒ'CIA, (δις twice, οικος a house) a separate class, and also an order of another class, in the artificial system of Linneus; characterized by the unisexual flowers of the same species being produced on distinct individuals.

DIŒ'CIOUS, *DIOI'CUS*, possessing the characteristic mentioned under Diœcea.

DIOSCOREA'CEÆ, DIOSCO'REÆ, (from the genus Dioscorea) The Yam tribe. A natural order of Dicotyledones.

DIPLECOLO'BEÆ, (δις twice, πλεκω to fold, λοϐος a lobe) a sub-order of Cruciferæ, characterized by an embryo, whose colytedons being incumbent on the radicle, are also twice folded, transversely, in the manner represented. Fig. 73.

73

DIPLO'TEGIS, *DIPLO'TEGIA*, *DIPLO'TEGIUM*, (διπλο ς double, τεγη a roof) a dry fruit, formed as the capsule, but from being "inferior" is also invested by the persistent calyx; as in Campanula.

DIPSA'CEÆ, (from the genus Dipsacus) the Scabious tribe. A natural order of Dicotyledones.

DIPTERA'CEÆ, DIPTEROCAR'PEÆ, (from the genus Dipterocarpus) the Camphor-tree tribe. A natural order of Dicotyledones.

DIP'TEROUS, *DIP'TERUS*. (δις twice, πτερον a wing) having two membranous expansions, termed wings; as the seeds of Halesia diptera.

Direct'e-venosus, (*Directe* simply, straight, *venosus* full of veins) synonyme for Digitinervius; straight ribbed.

Disappear'ing, where the branching of a tree or shrub is continued to a great extent, so that the trunk or leading stems appear to be subdivided to excess.

Discoi'd, Discoi'dal; *Discoi'deus*, (δισκος a disk, ειδος a resemblance) a round, somewhat thickened lamina, the margins of which are also rounded. Also used to designate a large spot of colour surrounded by some other colour.

Disk, (*Dis'cus* a quoit) certain fleshy expansions between the stamens and pistil, which occur in some flowers, and are considered to result from the abortion of an inner whorl of stamens. Applied also to that portion of the surface of the limb of a leaf which is included within the margin. Also, the central portion occupied by the flowers in a capitulum, umbel, or corymb. A synonyme, also, for the receptacle of Compositæ.

Dissec'tus, (cut in pieces) where the segments, as in some leaves, are very numerous and deeply cut.

Dissemina'tion, *Dissemina'tio*, (*Dissemino* to spread abroad) the manner in which the ripe seeds of plants are naturally dispersed.

Dissep'iment, *Dissepimen'tum*, (*Dissepio* to part, or separate) vertical planes in the interior of an ovary or pericarp, dividing it wholly or partially into two or more cells.

Dissi'liens, (Bursting asunder) when the valves of a seed vessel bursts with elasticity.

Dissim'ilar, (*Dissimilis* unlike) when similar organs assume different forms in the same individual; as some of the anthers in the genus Cassia.

Dis'tant, (*Distans*) when similar parts are not closely aggregated; used in opposition to "dense" or "approximate."

Dis'tichous, *Dis'tichus*, (δις twice, στιχος a rank) longitudinally arranged in two rows, on opposite sides of a common axis. Fig. 74.

74

Disti'nct, (*Distinctus* divided into parts) when any part or organ is wholly unconnected with those near it.

Distrac'tilis, (*Distractus* drawn asunder) applied to the connective, when it is so much enlarged as to keep the lobes of the anther wide apart; as in the genus Salvia.

Diur'nal, (*Diurnus* daily) used synonymously with ephemeral. Also, lasting during daylight.

Diva'ricate, (*Divaricatus* severed, straddling) branching off at a high angle, and spreading irregularly in various directions.

Diverg'ing, *Diver'gens*, (*De* from, *vergo* to incline) when similar parts, approximating at their bases, incline away from each other towards their summits; as the two follicles forming the fruit of Asclepias.

Divi'ded, (*Divisus*) where incisions, or indentations, extend nearly to the base.

Dodecagynia, (δωδεκα twelve, γυνη a woman) an order in the artificial system of Linneus, characterized by flowers which have twelve pistils.

Dodeca'gynous, *Dodeca'gynus*, possessing the structure of flowers in dodecagynia; or even including those which have only one pistil, provided there are twelve distinct styles, or stigmas.

Dodecandria (δωδεκα twelve, κνηρ a man) the eleventh class in the artificial system of Linneus, including flowers with twelve stamens; or rather those which have between twelve and twenty, provided they are not attached to the calyx.

Dodecan'drous, *Dodecan'der*, *Dodecan'dricus*, *Dodecan'drus*, having twelve stamens in the flowers. See Dodecandria.

Dodra'ns, (nine inches) a span; of about nine inches.

Dola'briform, *Dolabra'tus*, *Dolabrifor'mis*, (*Dola'bra* an axe, *forma* a form) synonyme for axe-shaped.

Dolia rius, *Dolia'tus*, (*Dolium* a tub) synonyme for *circinatus*.

Dor sal. *Dorsa'lis*, (*Dorsum* the back) attached to the back of any organ.

Dor'sum, the back.

Dotted, where spots or impressions of any kind are very small and numerous.

Double, (*Du'plex*) when applied to the entire flower, it signifies that monstrous condition in which the parts of the inner floral whorls, the stamens and carpels, become converted to petals. Applied to the calyx or corolla separately, it refers to certain examples in which these organs appear to consist of more than the usual or normal number of subordinate parts, and thus seem as if they were double. Also a synonyme for Twin.

DROOP'ING, synonyme for cernuous.

DROSERA'CEÆ, (from the genus Drosera) the Sundew tribe. A natural Order of Dicotyledones.

DRUPA'CEOUS, *DRUPA'CEUS*, either possessing the character of a Drupe, or resembling one in outward appearance.

DRU'PE, *DRU'PA*, (*DRUPÆ* unripe olives) a fruit, composed of an indehiscent, superior, one-celled pericarp, fleshy externally and bony within, containing one or two seeds. The stone fruits, as plums, apricots, &c. are examples.

DRU'PEL, *DRUPE'OLA*, *DRUPEL'LA*, a very small Drupe. The fruit of the blackberry and other Rubi is composed of several Drupellæ seated on a pulpy receptacle.

DUCT, *DUC'TUS*, (*DUCO* to conduct) a membranous tube, one of those which constitute the Vascular texture; with or without markings on the surface; but not accompanied in a spirally coiled fibre, as the Tracheæ. (See Ex. at figure 28.)

DUC'TUS-INTERCELLULA'RES, intercellular passages.

DUCTULO'SÆ, a section of Acotyledones, characterized by an imperfect vascular texture, composed of Ducts.

DUMETO'SUS, (*DUMETUM* a thicket, *DUMUS* a bush) having the character or appearance of a bush.

DUMO'SUS, full of bushes. Synonyme for Dumetosus.

DUPLEX, double.

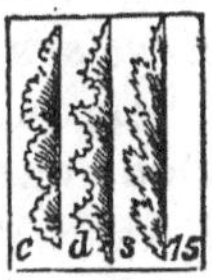

DUPLICA'TO- (*DUPLICATUS* doubled) when compounded with the words Crenate, Dentate, Serrate, it implies that the incisions on the margins of leaves bearing these names are themselves crenated, dentated, and serrated. Fig. 75, (c.) (d.) (s.) respectively.

c d s 75

DUPLICA'TO-PINNA'TUS, synonyme for Bi-pinnatus.

DUPLICA'TO-TERNA'TUS, synonyme for Bi-ternate. N.B.---In the explanation of this latter term, at page 27, the last word "primary," should have been "leaflets on a ternate leaf." See figure 76.

76

DUPLICA'TUS, (doubled) twin.

DURA'MEN, (*DURUS* hard) heart-wood.

29

Dus'ty, where an otherwise smooth surface is covered with minute granular incrustations, resembling dust.

Dwarf, of small size compared with other species of the same genus; or with other varieties of the same species.

Eared, see *Auricula*, and figure 34.

Ebena'ceæ, (*Ebenus* the Ebony tree) the Ebony tribe. A natural Order of Dicotyledones.

Ebori'nus, (*Eborius* made of ivory, *Eburneus* white like ivory) of the colour of ivory; white, slightly tinged with yellow, and with a tendency to a wavy lustre.

Ebractea'tus, (*e* without, *bractea* a bract) destitute of bracts.

Echina'tus (set with prickles) bristly. Applied to surfaces which are covered with "Bristles;" or to surfaces coated with straight "prickles."

Echinula'tus, (diminutive for *echinatus*) rough with small bristles, prickles, or tubercles.

Ectophlæodes, (εχτος without, φλοιὸς bark) living like some lichens, on the surface of other plants.

Edged, when any part, or patch of colour, is surrounded by a narrow rim of a different colour.

Effigura'tus, (*figu'ra* a figure) when the form of any part is completed by the full development of all its subordinate parts.

Efflorescen'tia, (*effloresco* to blossom or flower) the period at which a plant expands its flowers.

Effu'sio, (a pouring out) synonyme for "*expan'sio.*"

Effu'sus (poured out) synonyme for "*patulus*," also for "expansus."

Egg-shaped, synonyme for Ovate.

Ehretia'ceæ, (from the genus Ehretia) a group considered either as a natural Order of Dicotyledones: or as a tribe of the Order Boraginaceæ.

Elæagna'ceæ, Elæag'neæ, Elæag'ni, (from the genus Elæagnus) the Oleaster tribe, a natural Order of Dicotyledones.

Elæocarpa'ceæ, Elæocar'peæ, (from the genus Elæocarpus) a natural Order of Dicotyledones.

Ela'ter, (ελατηρ a driver) a membranous, elastic, and spirally twisted filament, of which there are several in the thecæ of some of the Hepaticæ. When the theca is ripe, the Elateres uncoil and assist in dispersing the sporules.

ELATE'RIUM, (ελατειρη that which driveth away) synonyme for "*Coccum*."

ELATINA'CEÆ, ELATI'NEÆ, (from the genus Elatine) the water-pepper tribe. A natural Order of Dicotyledones.

ELA'TUS, (Lofty) tall.

ELEMENTARY-ORGANS, the vesicles and tubes of which the Cellular and Vascular tissues are composed.

ELL, roughly estimated at two feet, or about the length of the arm.

ELLIPSOI'DAL, ELLIPSOI'DEUS, (Ελλειψις a figure in rhetoric; whence we derive the word "Ellipse," a geometric figure) approaching the form of an Ellipsoid; a solid figure formed by the revolution of an Ellipse about its major axis.

ELLIP'TICAL, *ELLIP'TICUS*. (from Ελλειφις, as above) approaching the form of an ellipse. Frequently, but inaccurately, used synonymously with "oval;" whereas an ellipse is necessarily rounded at the extremities, and therefore rather agreeing with oblong in this particular. It may be considered as an oval rounded at the ends, or as an oblong widened in its smaller diameter.

ELON'GATED, *ELONGA'TUS*, (*Longus* long, tall) when any part or organ is in any way remarkable for its length, in comparison with his breadth.

ELOCULA'RIS, (*E* without, *loculus* a partition) synonyme for Unilocularis.

ELITRI'CULUS, (diminutive, from ελυτρον a covering) synonyme for Flosculus.

EMAR'GINATE, *EMARGINA'TUS*, (*E* out of, *margo* the extremity or margin) slightly notched at the summit, as if a piece had been cut out. Fig. 77.

77

EMBRA'CING, where the base of an organ extends on each side partially round the part to which it is attached, as in the *AMPLEX'ICAUL* leaf. Fig. 20, page 14.

EM'BRYO, *EM'BRYO*, (εμϐρυον a fœtus) the rudiment of a plant contained in the seed. It makes its first appearance soon after the pollen has fertilized the ovule.

EMBRYONIC-SACK, an integument in the ovule, within which the embryo is developed.

EMBRYO'TEGA, *FMBRYO'TEGIUM*, (εμβρυον a fœtus, τεγη a covering) a callosity in the seed-covering of some seeds, situate near the hilum: it is detached by the protrusion of the radicle, in germination.

EMBRYO'TROPHA, (εμϐρυον a fœtus, τροφη nourishment). Synonyme for Perispermium. Also for Amnios.

EMER'SUS, (*EMER'GO* to swim) where the upper extremities of a plant, or of a leaf, rise above the water, the rest continuing submerged.

EMPETRA'CEÆ, EMPE'TREÆ, (from the genus Empetrum) the Crowberry tribe. A natural Order of Dicotyledones.

ENDECA'GYNOUS, *ENDECA'GYNUS*, (ενδεκα eleven, γυνη a woman) possessing eleven pistils.

ENDECAN'DROUS, *ENDECAN'DRUS*, (ενδεκα eleven, ανηρ a man) possessing eleven stamens. No flowers are strictly characterized by possessing either eleven stamens or eleven pistils, but as such conditions occur from accidental abortions or monstrous developments, these terms are in use.

ENDEM'IC, *ENDEM'ICUS*, (εν among, δημος the people) when the geographical range of any species, or natural group, is confined within the limits of a particular Botanical region.

EN'DOCARP, *ENDOCAR'PIUM*, (ενδον within, καρπος fruit) the inner coat of the pericarp, membranous in some cases, but in others hard and bony, as in stone fruits.

ENDO'GENÆ, synonyme for Monocotyledones. See Endogenous.

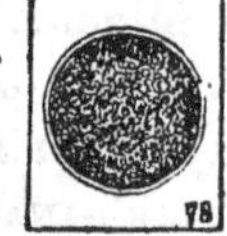
78

ENDO'GENOUS, *ENDO'GENUS*, (ενδον within, γενναω to produce) possessing the internal structure of Monocotyledones; viz. where the newest cellular and vascular tissue occurs within the older; the vessels are also scattered, and not collected in concentric rings, as in the stems of Dicotyledones. Fig. 78.

ENDOPHYL'LOUS, *ENDOPHYL'LUS*, (ἔνδον within, φυλλον a leaf,) used synonymously with Monocotyledones; on account of the manner in which the first leaves of Monocotyledones are evolved; viz. from within a sheath.

ENDOPLEU'RA, (ἔνδον within, πλευρὰ a side) the innermost of the integuments of the seed, immediately investing the embryo and albumen.

ENDOP'TILUS, (ἔνδον within, πτίλον a feather) synonyme for Endophyllus.

ENDORHI'ZÆ, synonyme for Monocotyledones. See Endorhizous.

ENDORHI'ZOUS, *ENDORHI'ZUS*, (ἔνδον within, ῥιζα a root) used synonymously with Monocotyledonous; because, in the germination of Monocotyledones, the radicle, instead of elongating, is burst at its apex or sides, by secondary radicles or fibres, which are then protruded through the openings.

ENDOSMO'SE, *ENDOSMO'SIS*, (ἔνδον within, ωσμος impulsion) that property of membranous tissue by which fluids of unequal densities, when placed on opposite sides of it, are enabled to pass through and intermix.

ENDOSPE'RM, *ENDOSPER'MIUM*, (ἔνδον within, σπέρμα a seed) synonyme for perisperm, or albumen.

ENDOSTO'ME, *ENDOS'TOMA*, (ενδον within, στόμα a mouth) the perforation in the secundine or tegmen, constituting the innermost of the perforations which together make up the foramen.

ENDOTHE'CIUM, (ενδον within, θηκη a box) the inner lining of the anther-cells.

ENE'ILEMA, (εν within, εἰλέω to shut up) synonyme for the inner skin of the seed.

ENER'VIS, *ENER'VIUS*, (*E* without, *NERVUS* a nerve) without nerves or veins.

ENNEAGY'NIA, (ἐννέα nine, γυνή a woman) an artificial Order, characterized by flowers with nine free pistils, styles, or stigmas.

ENNEAN'DRIA, (ἐννεά nine, ανηρ a man) the ninth class in the artificial system of Linneus, containing a few plants which have nine stamens in each flower.

ENNEAN'DROUS, *ENNEAN'DER*, *ENNEAN'DRICUS*, where a flower has nine stamens, as in the class Enneandria.

ENNEAPE'TALUS, (εννεά nine, πεταλον a petal) having nine petals.

ENO'DIS, (*E* without, *NODUS* a knot) where a stem, or other part, is without joints or knots.

EN'SIFORM, *ENSIFOR'MIS*, (*ENSIS* a sword, *FORMA* a shape) synonyme for "Sword-shaped."

ENTAN'GLED, irregularly or confusedly interlaced; as in the case of some branches; or in some fibres of the root; or of the hairy pubescence of some plants.

ENTIRE, without any traces of division, incision, or separation. Sometimes used synonymously with "simple," or very "slightly divided," in contradistinction to "compound" or "deeply incised."

ENTODIS'CALIS, (εντος within, δισκος a disk) inserted, as some stamens, within-side a disk.

ENVEL'OPING, synonyme for involute.

EPACRIDA'CEÆ, EPACRI'DEÆ, (from the genus Epacris) a natural Order of Dicotyledones. Otherwise considered a group subordinate to Ericaceæ.

EPHEM'ERAL, EPHEM'EROUS, *EPHEM'ERUS*, (επι for, ημερα a day) existing for, or less than, one day. As where a corolla expands for a few hours at most, and then fades.

EPIBLAS'TUS, *EPIBLAS'TANUS*, (επι upon, βλαστος a bud) an unguiform appendage, seated on the anterior part of the plumule (*BLASTUS*) of certain Graminaceæ; and considered to be either a second cotyledon, a prolongation of the lower part of the plumule itself, or else of the upper portion of the radicle. See figure 79. (*e*).

e

79

EPICAL'YCIUS, (επι upon, καλυξ the Calyx) synonymous with Epistamineus.

E'PICARP, *EPICAR'PIUM*, (επι upon, καρπος fruit) the outer skin or coat of the pericarp, when ripened into a fruit.

EPICARPAN'THUS, *EPICAR'PIUS*, *EPICAR'PICUS*, (επι upon, καρπος a fruit, ανθος a flower) synonyme for "superior," as applied to a flower, or to the several parts of a flower.

EPICHI'LIUM, (επι upon, χεῖλος a lip) the upper portion of the lip of any Orchidaceous plant, when this organ is divided into two parts which are dissimilar in appearance.

EPICLI'NUS, (επι upon, κλινή a bed) attached to the receptacle of the flower; as the nectary in Labiatæ.

EPICOROLLA'TUS, (επι upon, *COROLLA* the corolla) having an epigynous corolla.

EPIDER'MIS, *EPIDER'MA*, *EPIDER'MIS*, (επιδερμις the cuticle) a delicate membrane coating the surface of the various parts of plants, composed of flattened vesicles of the cellular tissue.

EPIGÆUS (επι upon, γῆ the earth) growing on land, in contradiction to growing in the water; also, when any part of a terrestrial plant grows close to the earth.

Epi'gynous, *Epi'gynus*, (επι upon, γυνη a woman) where the outer whorls of the flower adhere to the ovary, so that their upper portions alone are free, and thus they appear to be seated on it, or to arise from it in the manner termed "superior," as in the Umbelliferæ. Fig. 80.

Epilobiaceæ, (from the genus Epilobium) synonyme for Onagraceæ.

Epine'ma, (επὶ upon, νημα a thread) the superior portion of the filament in Compositæ, bearing the anther.

Epiphrag'ma, (επὶ upon, φραγμα a division) a membrane which closes the orifice of the theca in some mosses, and remains after the lid has fallen off.

Epiph'yllous, *Epiphyl'lus* (επὶ upon, φυλλον) either growing upon, or inserted on the leaf.

Epiphy'te, *Epi'phyton*, (επὶ upon, φυτον a plant) certain aerial plants, (as some Mosses, Lichens, Orchidaceæ, &c.) which attach themselves to others for support, but derive no nutriment from them, as the true parasites do.

Epiptera'tus, (επὶ upon, πτερον a wing) when any part is prolonged in the form of a thin expansion termed a wing.

Epirrheo'logy, *Epirreolo'gia*, (επιρρέω to overflow, λογος a discourse) the department of Botanical Physiology which treats of the effects of external agents on living plants.

Epispe'rm, *Episper'mium*, (επὶ upon, σπερμα the seed) the seed-cover; used synonymously with Lorica, Perisperm, and Spermodermis.

E'qual, (*Æqualis*, *Æquans*) where one part is of the same general form, disposition, and size, as some other part with which it is compared. Used also synonymously with "regular."

E'qually-pinna'te, synonyme for "abruptly-pinnate." See figure 3.

E'qual-si'ded, synonyme for "equal," when applied to the two sides of any particular organ.

Equinoc'tial, (*Æquinoctia'lis*,) plants whose flowers expand and close at particular hours of the day,

Equiseta'ceæ, (from the genus Equisetum) the Horse-tail tribe. A natural Order of Acotyledones.

E'QUITANT, *EQUITATI'VUS*, (*EQUITANS* riding) a form of vernation, in which the leaves are folded forwards longitudinally on the midrib, so that their edges meet, and each embraces the one which is placed next within it—a transverse section of such arrangement is represented at figure 81.

ERECT', (*ERECTUS*) when any part or organ stands perpendicularly, or very nearly so, to the surface to which its base is attached.

ERIAN'THUS, (εριον wool, ανθος a flower) when some parts of a flower are covered with a woolly or cottony pubescence.

ERICA'CEÆ, ERI'CÆ, ERI'CEÆ, ERICI'NEÆ, (from the genus Erica) the Heath tribe. A natural Order of Dicotyledones.

ERICE'TINUS, (*ERICÆUS* found upon heaths) growing on heaths.

ERIOCAU'LEÆ, (from the genus Eriocaulon) synonyme for Restiaceæ.

ERIO'PHORUS, (εριον wool, φερω to bear) covered with woolly or downy pubescence.

ERO'SE, (*ERO'SUS* gnawn round-a-bout) synonyme for "gnawed."

ERYTHROS'TOMUM, (ερυθρὸς red, στόμα the mouth) a synonyme for "Etærio."

ESCALLONIA'CEÆ, ESCALLO'NIEÆ, (from the genus Escallonia) a natural Order of Dicotyledones; otherwise considered a tribe subordinate to Saxifragaceæ.

ESSEN'TIAL, *ESSENTIA'LIS*, (*ESSENTIA* the essence of anything) the most prominent characteristics by which a particular species, or a particular group of plants is separated from all others.

ETÆ'RIO, *ETA'IRIUM*, (εταιρια a society) a fruit composed of several distinct one-seeded pericarps, (akenia and caryopses of different authors, or drupellæ) arranged upon an elevated receptacle or torus, which may be either dry or fleshy. This definition embraces the three modifications presented respectively by Ranunculus, Fragaria, and Rubus.

ETIOLA'TED, *ETIOLA'TUS*, the effect of blanching the leaves; and lengthening the stem, when a plant is suffered to grow in the dark, or in a much obscured situation.

EUPHORBIA'CEÆ, EUPHOR'BIÆ (from the genus Euphorbia) the Euphorbium tribe. A natural Order of Dicotyledones.

Evalvis, (*e* without, *valva* a valve) synonyme for "indehiscens."

Evanescente-venosus, (*Evanescens* vanishing, *venosus* full of veins) when the lateral veins of a leaf do not extend so far as the margin.

E'ven, where a surface is without inequalities of any description.

Exalbu'minous, ***Exalbumino'sus***, (*Ex* without, *albumen*) a seed which has no distinct albumen, or none but what is contained within the cotyledons themselves.

Exaspera'tus, (sharpened) rough.

Ex'cipulus, ***Exci'pula***, wort-like excrescences on the thallus of some lichens, pierced with a narrow opening. The portion of the thallus which forms the rim round the base of Apothecia.

Excitabil'ity, ***Excitabil'itas***, (*Excito* to stir or move,) that faculty by which living beings take cognizance of external stimuli, and obey their influence. This is considered by some vegetable physiologists to be the sole vital property distinguishable in plants.

Excre'tion, (*Excretio* the rejection of excrement) the action by which a superabundance of secreted matter is rejected from a secreting vessel. Also the matter itself thus excreted: gum, resin, &c. are examples.

Excur'rent, (*Excurrens* sallying forth) protruding beyond the usual limits; as where the nerve in the Moss-leaf is extended beyond the apex in the form of a point or bristle.

Exhala'tion, ***Exhala'tio***, a vital function by which the stomata are made to discharge a large portion (about two-thirds) of the water introduced by absorption through the spongioies.

Ex'ogens, **Exo'genæ**, used synonymously with Docotyledones, because the stems of such plants have an exogenous structure.

Exo'genous, ***Exo'genus***, (ἔξω outwards, γεννάω to beget) the peculiar structure of Dicotyledonous stems; where the successive deposits of newly organized wood are exterior to the old ones. They consist, ultimately, of concentric layers of wood (w) surrounding a central pith (p); concentric layers of bark (d), the newest being the innermost, surrounding the wood. Both these are intersected at right angles by vertical plates

82

of cellular tissue, composing the medullary rays, (*r*), Fig. 82.

Exo'gynus, (ἐξω outwards, γυνη a woman) where the style is exserted beyond the flower.

Exo'phyllous, *Exo'phyllus*, (ἐξω without, φυλλον a leaf) not having a foliaceous sheath. Used synonymously with Exorhizous, because the cotyledons of such plants have no coleoptile like the Endorhizous.

Exop'tile, *Exop'tilus*, (εξω without, πτίλον a wing) synonyme for Exophyllous.

Exorhi'zæ, used synonymously with Dicotyledones, to express the exorhizal development of the radical in germination.

Exorhi'zal, *Exorhi'zus*, (εξω outside, ριζα a root) the peculiar mode in which the radicle of dicotyledones is developed in germination; elongating at once from the radicular extremity of the embryo, and not bursting through an outer coat, as in Monocotyledones.

Exosmo'se, *Exosmo'sis*, (εξω outwards, ὤσμος an impulsion) the effect opposed to Endosmose, referring to the current which passes from within outwards.

Exosto'me, *Exos'toma*, (εξω without, στομα a mouth) the perforation in the primine or testa (the outermost covering of the nucleus) which, together with the endostome, completes the foramen.

Exosto'sis, (ἐξόστουσις a bony protuberance) a wart-like excrescence, many of which are developed on the roots of several Leguminosæ.

Exothe'cium (εξω without, θηκη a case) the outer coat of the anther.

Expand'ed, (*Expansus* spread out) when the flower is fully blown. A synonyme for "Diffuse."

Explana'tus, (made smooth) spread out flat, as the limb of the corolla in many monopetalous flowers.

Exculp'tus, (*ex* out of, *sculpo* to engrave) where there exists a small depression, as though a piece had been cut out; as in the seeds of Anchusa.

Exsert'ed, (*Exsertus* thrust out) when one part protrudes beyond another by which it is surrounded; as the stamens or style beyond the mouth of some tubular corollas.

Exten'sus (stretched out) synonyme for "Diffuse."

Exstipula'tus, (*Ex* without, *stipula* a stipule) destitute of stipules.

EXTE'RIOR, EXTER'NAL, (*EXTERNUS* outward) exposed, and not invested by any part or covering.

EXTENUA'TUS, (made thin) synonyme for *VIRGATUS*.

EXTEN'SUS, (stretched out) synonyme for *DILATATUS*.

EXTRA-AXILLA'RIS, (*EXTRA* outside, *AXILLA* the axil) when a bud, instead of being placed in the axil of the leaf, is developed above or on one side of it.

EXTRO'RSE, *EXTROR'SUS*, (*EXTRA* externally, *ORSUS* originating) when the slit through which the pollen escapes from the anther is towards the outside of the flower, and not, as is more usual, towards the pistil.

FABA'CEÆ, (*FABA* a bean) synonyme for Leguminosæ.

FA'CIES, (a face) the general habit or appearance assumed by each particular species.

FACTI'TIOUS, *FACTI'TIUS*, "Artificial."

FADING, withering up without falling off; at least not for some time after flowering.

FÆCULA, (dregs of wine) the farinaceous matter which forms starch, &c.

FAL'CATE, *FALCA'RIUS*, *FALCATO'RIUS*, *FALCIFOR'MIS*, (*FALCA'TUS* hooked) plane and curved, with the edges parallel. Fig. 83.

83

FALSINER'VIS, (*FALSUS* false, *NERVUS* a nerve) when the nerves of a leaf are formed of elongated cellular tissue, without any vessels, as in Cryptogamic plants.

FALSE, (*FALSUS*) where there exists a close resemblance to some particular structure, but which resemblance has originated in some unusual and irregular manner. Ex. gr. The spurious cells sometimes formed in a legumen (as in that of the common bean) by a development of the cellular tissue of the endocarp, and not by an actual dissepiment. Those which occur in the fruit of Nigella, by the unusual separation of the epicarp and mesocarp, and inflation of the intermediate space.

FAM'ILY, (*FAMILIA*) a synonyme for "Order."

FAN-SHAPED, synonyme for Flabelliform.

FARC'TUS, (stuffed) used in contradistinction to hollow or tubular.

FARINA'CEUS, *FARINO'SUS*, (*FARI'NA* meal) mealy.

FASCIA'LIS, *FASCIOLA'RIS*, *FASCIOLA'TUS*, synonymes for *FASCIATUS*.

FAS'CIATED, (*FASCIATUS* swathed) when contiguous parts are unusually grafted and grown together; as some stems and branches, which then assume a flattened instead of a rounded appearance. Used also synonymously with "Banded," and "Band-shaped," (*FASCIARIUS*).

FASCI'CLED, *FASCI'CULATE*, *FASCICULA'RIS*, *FASCICULA'TUS*, (*FASCICULA* a little bundle) where several similar parts originate at the same spot, and are collected as it were, into a bundle.

FASTI'GIATE, (*FASTIGIATUS* sharpened at the top like a pyramid) where many like parts are parallel, and point upwards; as the branches of Populus fastigiata.

FAUX, (the gorge) the throat.

FAVEOLA'TUS, *FAVO'SUS*, *FAVULO'SUS*, (*FAVUS* the honeycomb.) synonyme for Alveolate.

FEATHER-VEINED, synonyme for Penninerved.

FEATHERY, synonyme for Plumose.

FEC'ULA, see *FÆCULA*.

FEMI'NEUS, (*FEMINA* a woman) containing a pistil, but no stamens.

FENESTRA'LIS, (*FENESTRA'TUS* having windows) pierced with holes or openings of somewhat considerable dimensions.

FERRU'GINOUS, *FERRUGINO'SUS*, (*FERRUGINEUS* of the colour of rusty iron) red mixed with much grey.

FER'TILE, (*FERTILIS*) producing fruit. Also, capable of effecting the process of fertilization; as the anthers when filled with pollen.

FERTILIZA'TION, *FERTILIZA'TIO*, (*FERTILIS* fertile) the reproductive function by which the action of the pollen renders the ovule fertile.

FI'BRE, (*FIBRA* a filament) extremely fine and transparent hair-like condition of the elementary vegetable texture, which, together with membrane, enters into the composition of several forms of tissue.

FI'BRIL, *FIBRIL'LA*, (diminutive for Fibra) a fine ultimate hair-like subdivision of the root; or hair-like appendages to its branches. The roots of Lichens are termed Fibrillæ.

FIBRILLA'TUS, possessing Fibrillæ.

FI'BROUS, *FIBRO'SUS*, (*FIBRA* a filament) consisting of many thread-like portions; as the root of an Onion. Or possessing a structure separable into woody fibres, as the outer coat of the Cocoa-nut.

FICOI'DEÆ, a natural Order of Dicotyledones.

FID'DLE-SHAPED, synonyme for "Panduriform."

FIL'AMENT, *FILAMEN'TUM*, (*FILUM* a thread) the stalk which in many stamens supports the anther.

FILAMEN'TOUS, *FILAMENTO'SUS*, (*FILUM* a thread) composed of thread-like bodies; as Conferva. Or bearing filaments; as the leaves of Yucca filamentosa.

FILA'TUS, (*FILUM* a thread) synonyme for Virgatus.

FI'LICES, (*FILIX* a fern) the Fern tribe. An extensive group of Acotyledones, sometimes considered as a single Order; but now more usually subdivided into several distinct Orders.

FIL'IFORM, *FILIFOR'MIS*, (*FILUM* a thread, *FORMA* shape) cylindrical and slender, like a thread.

FILIPEN'DULOUS, *FILIPEN'DULUS*, (*FILUM* a thread, *PENDULUS* hanging down) where tuberous swellings are developed in the middle or at the extremities of filiform rootlets; as in Spiræa filipendula.

FIMBRIA'TUS, fringed.

FIN'GERED, synonyme for Digitate.

FIS'SUS, (cleft) split.

FIS'TULA-SPI'RALIS, (*FISTULA* a pipe, *SPIRALIS* spiral) synonyme for Trachea.

FIS'TULAR, FIS'TULOSE, FIS'TULOUS, (*FISTULOSUS* hollow) cylindrical and hollow; and either with transverse diaphragms, as in the stems of some Junci, or without them, as in the stems and leaves of the Onion.

FLABEL'LIFORM, *FLABELLIFOR'MIS*, *FLABELLA'TUS*, (*FLABELLUM* a fan, *FORMA* shape) shaped, and sometimes plaited, like a fan; rounded at the summit and cuneate at the base. Fig. 84.

84

FLAC'CID, (*FLACCIDUS* withered, weak) bending without elasticity; as some peduncles under the weight of flowers.

FLACOURTIA'CEÆ, FLACOURTIA'NEÆ, (from the genus Flacourtia) a natural Order of Dicotyledones.

FLAGEL'LUM, (a whip) synonyme for Ramulus; also for Sarmentum.

FLAGEL'LIFORM, *FLAGELLIFOR'MIS*, *FLAGELLA'RIS*, (*FLAGEL'LUM* a whip, *FORMA* a shape) flexible, narrow, and tapering, like the thong of a whip. Ex. gr. The stems of Clematis vitalba.

FLAM'MEUS, (of a flame colour) brilliant red.

Flaves'cens, *Fla'vus*, *Fla'vidus*, (varieties of yellow) pale yellow, or pure yellow diluted.

Flavovi'rens, (*Flavus* yellow, *virens* green) yellowish green. Yellow with a little blue.

Flesh'y, when the flesh is firm and succulent.

Flex'ible, (*Flexilis*, *Flexibilis*) capable of being bent, but returning with elasticity to its original state.

Flexuo'se, (*Flexuosus* full of turnings) bending gently to and fro in opposite directions.

Float'ing, synonyme for Swimming. Also used where one part lies on the surface of the water, but is united to another which is submerged, and even attached to the ground; as in several species of Potamogeton.

Floc'cus, (a lock of wool) one of the numerous filaments intermixed with the sporules of some lichens. Also, the filaments of which Byssaceæ are composed.

Flocco'se, *Flocco'sus*, (*Floccus* a lock of wool) when dense hairy pubescence falls off in little tufts.

Flo'ra, (the goddess of flowers) the aggregate of all the species of plants inhabiting a particular country.

Flo'ral, *Flora'lis*, belonging to the flower; or, seated about the flower-stalk and near the flower.

Flo'ret, *Flo'rula*, (diminutive of *flos* a flower) one of the little flowers in a head; as in Compositæ.

Flos'culous, *Flosculo'sus*, (*Flosculus* a little flower) when the corolla of a floret is tubular.

Flos'culus, (a little flower) a floret.

Flow'er, (*Flos*) the apparatus destined for the production of seed, and necessarily including one or other, or both, of the sexual organs.

Flow'er-bud, the assemblage of the various parts composing the flower, previous to their expansion.

Flu'itans floating.

Flumina'lis, (*Flumineus*, *Fluvialis*, belonging to a river) applied to plants which grow in running streams.

Fluvia'les, (*Fluvialis* belonging to a river) a natural Order of Monocotyledones, of which some of the species grow in fresh, and others in salt water.

Folia'ceous, (*Foliaceus* of or like leaves) of the nature of a leaf. Or resembling the more usual character of leaves in being thin and membranous.

Folia'ris, (*Folium* a leaf) inserted on or forming an appendage to, the leaf. Synonyme for Epiphyllus."

Folia'tion, *Folia'tio*, (*Folium* a leaf) the period when the leaf-buds begin to expand. Also a synonyme for Vernation.

Foliifor'mis, (*Folium* a leaf, *forma* a shape) used synonymously with *foliaceus*.

Folio'sus, (*Folium* a leaf) when the leaves are particularly numerous on a plant.

Fol'licle, (*Folliculus* a little leathern bag) an univalved inflated pericarp, opening by a suture along one of the sides to which the seeds are attached. As in Colutea. Fig. 85.

Folli'cular, *Follicula'ris*, *Folliculifor'mis*, (*Folliculus* a little leathern bag, *forma* a shape) having the shape of a Follicle.

Foot, about twelve inches, roughly estimated at the size of a man's foot.

Fora'men, (a hole) the hole in the outer integuments of the ovule, through which the apex of the nucleus protrudes in the earlier stages of its development, until a short time after the process of fertilization has taken place. Fig. 86, (*f.*)

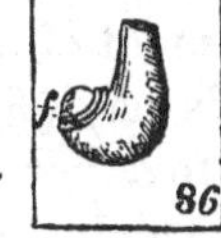

Forked, *Forcipa'tus*, (*Forceps* nippers) *Forfica'tus*, (*Forfex* scissors) separating into two distinct branches, more or less apart.

Fornica'tus, (arched over) when little scale-like appendages at the base of the limb of the corollas over-arch the mouth of the tube; as in Cynoglossum.

Fouquiera'ceæ, (from the genus Fouquiera) a natural Order of Dicotyledones.

Foveola'tus, (*Fovea* a pit-fall) impressed with small holes or depressions.

Fovil'la, (*Foveo* to nourish) the matter contained in the grains of pollen, consisting of minute granules floating in a liquid.

Fox'glove-shaped, a nearly cylindrical but somewhat irregular and inflated tube, formed like the corolla of a Digitalis.

Francoa'ceæ, (from the genus Francoa) a small natural Order of Dicotyledones.

Frankenia'ceæ, (from the genus Frankenia) a small natural Order of Dicotyledones.

FRIN'GED, when the margin is clothed with hair like appendages, or ciliæ.

FRONDESCEN'TIA, (*FRONS* a leaf) a synonyme for *VERNATIO*.

FRONDO'SUS, (full of leaves) furnished with leaves. Or, assuming a leaf-like condition; as the several parts of proliferous flowers.

FRON'DULA, (diminutive of *FRONS*, a leaf) a division in a Frond.

FROND, (*FRONS* a leaf) the foliage of Ferns; the foliaceous expansions of Hepaticæ, and others of the lower tubes of Acotyledones. The term has been further extended by some, to the foliage of Palms.

FROSTED, when a surface is covered with minute bladdery elevations of the parenchyma, which gives it the appearance of hoar-frost.

FRUCTIFICA'TION, *FRUCTIFICA'TIO*, (*FRUCTUS* fruit) the phenomena which attend the development of the fruit from its first appearance to maturity. The distribution or arrangement of the fruit itself on any plant.

FRUIT, (*FRUCTUS*) the matured pericarp and its contents, together with any external appendages of the inflorescence which may accompany them, and seem to form an integral part with them. Thus the calyx in the apple, the involucral bracts in the oak, the receptacle in the strawberry, are considered as forming parts of their respective fruit.

FRUMENTA'CEOUS, *FRUMENTA'CEUS*, (*FRUMENTUM* corn) producing farina or starch in sufficient abundance to be cultivated for economical purposes.

FRUSTRA'NEA, (*FRUSTRA* in vain) an Order of the Linnean Class, Syngenesia, characterized by having the flowers of the disk hermaphrodite, whilst those of the ray are either completely neuter, or else the pistil has no stigma and consequently cannot perfect its fruit.

FRUTES'CENT, (*FRUTESCENS*, *FRUTICANS* shooting up, *FRUTEX* a shrub) becoming a shrub; or having the appearance of a shrub.

FRU'TEX, a shrub.

FRUTI'CULUS, (diminutive of *FRUTEX*) a little shrub.

FUGA'CIOUS, (*FUGAX* swift, fading) soon falling off, or perishing; as does the calyx of poppies, even before the corolla is expanded.

FUL'CRUM, (a prop) a general name given to several of the appendages of the stem or leaves, which serve either for support or defence; as the prickle, tendril, hair, stipule, &c.

FULIGINO'SUS, (*FULIGINEUS* sooty) intense brown passing to black. The deepest grey with a little red.

FUL'VUS, (Tawny) orange-yellow with grey.

FUMARIA'CEÆ, FUMA'RIEÆ, (from the genus Fumaria) the Fumitory tribe. A natural group of Dicotyledones, considered either as a distinct Order, or as a tribe subordinate to Papaveraceæ.

FU'MEUS, *FUMO'SUS*, (Smoky) grey with a little red.

FUNA'LIS, (belonging to cords) synonyme for Funiliformis.

FU'NCTION, (*FUNCTIO*) the peculiar action induced by the agency of vitality upon any part of a living plant, when placed under the influence of certain stimuli.

FUNDAMEN'TAL-OR'GANS, the nutritive organs absolutely essential to the existence of the individual.

FUN'DUS-PLAN'TÆ, (*FUNDUS* a foundation) synonyme for "Collum."

FUN'GI, (*FUNGUS* a mushroom) the Mushroom tribe. An extensive Order, or rather Sub-Class, of Acotyledones.

FUN'GIFORM, *FUNGIFOR'MIS*, *FUNGILLIFOR'MIS*, (*FUNGUS* a mushroom, *FORMA* a shape) cylindrical, with the summit convex and capitate, like the pileus of an Agaric.

FUNGO'SUS, (Spongy) of a thick, coriaceous, and elastic substance.

FUNIC'ULAR-CHORD, (*FUNICULUS* a little rope) a chord-like appendage (*a*), by the intervention of which, in many cases, the seeds are attached, instead of being seated immediately on the placenta, (*p.*) As in Cruciferæ. Fig. 87.

FUNI'LIFORM, (*FUNIS* a rope, *FORMA* shape) tough, cylindrical, and flexible, like a chord; as the roots of arborescent monocotyledones.

FUN'NEL-SHA'PED, synonyme for Infundibuliform.

FURCA'TUS, *FURCELLA'TUS*, (*FURCA* a fork) forked.

FURFURA'CEUS, (*FURFUR* bran, scurf) covered with a meal-like powder.

FUR'ROWED, synonyme for Sulcate.

FUS'CUS, (Brown) brown with a grey tinge. Deep grey nd red.

Fu'siform, *Fusifor'mis*, *Fu'sinus*, (*Fusus* a spindle) a solid, whose transverse sections perpendicular to the axis are circular, and which tapers gradually at each end. As the root of a radish.

Galbu'lus, (the fruit of the Cyprus) a modification of the Cone; where the apex of each carpellary scale is much enlarged (as in Cupressus and Thuja) or even fleshy (as in Juniperus); so that collectively they form a rounded compact fruit. Fig. 88.

Ga'leate, (*Galea'tus* wearing a helmet) where a petal or other membranous organ is shaped in a hollow vaulted manner, like a helmet. As in Aconitum.

Galeifor'mis, (*Galea* a helmet, *forma* shape) synonyme for Galeate.

Galia'ceæ, from the genus Galium) synonyme for Stellatæ.

Gamopet'alous, *Gamopet'alus*, (γαμος marriage, πεταλον a leaf, as applied to a petal) synonyme for Monopetalous; on the supposition that the corollæ of such are formed by the union or grafting together of several petals.

Gamose'palous, *Gamose'palus*, (γαμος wedding, *sepalum* adopted word for a sepal) synonyme for Monosepalus; supposing such a calyx to be formed by the union of several sepals.

Garrya'ceæ, (from the genus Garrya) a natural Order of Dicotyledones, including only the single genus from which it takes its name.

Gasteromyce'tes, Gastero'myci, (γαστηρ the belly, μυκης a fungus) an extensive natural group, or Sub-Order, of Fungi.

Gela'tinous, *Gelatino'sus*, (*Gelatio* a freezing or congealing) having the consistence or general appearance of Jelly; as several of the simpler forms of Algæ.

Gem'inate, (*Gem'inatus* doubled) growing in pairs. Synonyme for Binate.

Gemma, (a young bud) a bud.

Gemma'tion, *Gemma'tio*, (*Gemma* a young bud) either, the disposition of the buds on plants; or, the period of their expansion.

Gem'mule, *Gem'mula*, (diminutive from *Gemma*) synonyme for Leaf-bud. The buds of Mosses. The reproductive corpuscles of Algæ.

GEN'ERAL, (*GENERALIS*) when an organ of a particular description invests certain parts of a plant, each of which bears an organ of a similar description; thus, the involucrum (*g*) at the base of a compound Umbel, as well as the Umbel itself, is termed "general," whilst those (*p*) at the base of the separate little umbels at the ends of the rays, and these umbels themselves, are termed partial. Fig. 89.

GENIC'ULATE, (*GENICULATUS* jointed) where any part is bent abruptly, so as to form a decided angle.

GENIC'ULUM, (a little knee, a joint) a Node.

GENS, (a nation) synonyme for Tribus.

GENTIANA'CEÆ, GENTIA'NEÆ, (from the genus Gentiana) the Gentian tribe. A natural Order of Dicotyledones.

GE'NUS, (*GENUS* a race) the smallest natural group composed of distinct species.

GEOBLAS'TUS, (γῆ the earth, βλαστάνω to sprout) an embryo whose cotyledons remain under ground during the process of germination; as in the common Pea.

GERANIA'CEÆ, GERA'NIÆ, (from the genus Geranium) the Geranium tribe. A natural Order of Dicotyledones.

GER'MEN, (a branch or bud) synonyme for *OVARIUM*.

GERMINA'TION, (*GERMINATIO*, *GERMINATUS*, a budding) the act, with its accompanying phenomena, by which seeds begin to grow, when they are placed under the conditions requisite to excite the vital energies of the dormant embryo.

GESNERA'CEÆ, GESNE'REÆ, GESNERIA'CEÆ, GESNERI'Æ, (from the genus Gesnera) a natural Order of Dicotyledones.

GIB'BOUS, *GIBBO'SUS*, (*GIBBUS* a swelling) where a part is convex, as though it were swollen, like the tube of the corolla in Antirrhinum majus; but more correctly applicable to solid parts which are convex.

GIGAN'TIC, (*GIGANTEUS* giant-like) when the dimensions of a particular species considerably exceed those of any of its congeners.

GILLIESIA'CEÆ, GILLIESI'EÆ, (from the genus Gilliesia) a small natural Order of Monocotyledones.

GILLS, vertical plates descending from the under side of the cap of an agaric, and radiating from the stipes. They form the Hymenium, or part in which the sporules lie. Fig. 90.

Gil'vus, (carnation, flesh-colour, or ashen grey) dirty yellow with a tinge of red. Orange-yellow and grey.

Githagi'neus, greenish red.

Gla'brous, *Glabra'tus*, (*Glaber* smooth) a surface wholly destitute of pubescence.

Gla'diate, *Gladia'tus*, (*Gladius* a sword) flat, straight, or slightly curved, with the edges parallel and apex acute; as the leaves of an Iris. Also a synonyme for "ancipital."

Gland, (*Glandula*) collections of cellular tissue, somewhat modified in its texture, and serving the purpose of a secreting organ. Some glands are sunk in the texture of the plants; others are elevated on pedicels, hair, &c.

Glanda'ceus, (*Glans* an acorn, chesnut, &c.) red-brown mixed with yellow. Yellowish-red and much grey.

Glan'dular, *Glandulo'sus*, (*Glandula* a gland) furnished with glands.

Glans, (a mast of Oak or other tree) a one- or few-seeded, dry, inferior, indehiscent pericarp, seated within a cupulary involucrum; as in the Oak, Nut, Chesnut, &c.

Glau'cous, *Glauces'cens*, (*Glaucus* sky-blue, sea-green, or fiery red) dull green with a very peculiar whitish blue lustre. Also, frosted with bloom of a bluish-green tinge.

Gleichenia'ceæ, Gleiche'neæ, (from the genus Gleichenia) a group of Ferns, either considered as a distinct natural Order, or as a Sub-order of the whole family, "Filices."

Glit'tering, where the lustre from a polished surface is not perfectly uniform.

Globo'se, (*Globosus*) nearly sphærical.

Glo'bular, *Globula'ris*, nearly sphærical.

Glo'bule, (*Globulus*) one of the two organs which constitute the fructification of Characeæ; viz the spherical body filled with elastic filaments. Fig. 91, (*g*); (*n*) is the other body or nucule.

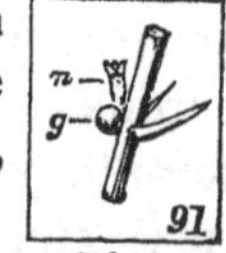

Globularia'ceæ, Globulari'neæ, from the genus Globularia) a natural Order of Dicotyledones, containing only the genus Globularia.

Globuli'ne, *Globuli'na*, (*Globulus* a globule) round transparent granules, formed in the cellular tissue, which constitute fecula.

Globulo'sus, synonyme for Globosus.

Glo'bulus, (a globule) globuline. Also used for a rounded

form of "shield" in some Lichens. Synonyme for Soredium.

GLO'CHIS, (γλωχις a point) a barb.

GLOCHI'DEUS, *GLOCHIDIA'TUS*, where the pubescence is formed of barbed bristles.

GLOMERA'TUS, (heaped up) synonyme for "*AGGLOMERATUS*."

GLOME'RULIS, (*GLOMERO* to heap up) synonyme for Soredium; synonyme for Capitulum.

GLO'MUS, synonyme for capitulum.

GLOSSOL'OGY, *GLOSSOLO'GIA*, (γωσσα the tongue, λογος a discourse). That department of Botany which contains an explanation of the technical terms employed in this science.

GLUE, a viscid secretion on the surface of some plants.

GLUME, (*GLUMA* chaff) the floral envelopes of grasses; but now more especially retained for the outermost husks only, which invest one or more flowers, composing the separate spikelets.

GLUMEL'LA, *GLUMEL'LULA*, (diminutive of gluma), in grasses, an innermost scale-like envelope to the ovarium; synonymous with Lodicula.

GLUMA'CEOUS, *GLUMA'CEUS*, GLUMO'SE, *GLUMO'SUS*, (*GLUMA* chaff) resembling the dry scale-like glumes of grasses; as the sepals of Junci.

GLU'TEN, (*GLU'TEN* glue) a viscid tenacious substance obtained from flour after washing away all the starch. Synonyme for Glue.

GNAW'ED, where the margin of a leaf or other membranaceous expansion is irregularly jagged, as though it had been bitten by a caterpillar.

GNETA'CEÆ, GNE'TEÆ, (from the genus Gnetum) a natural Order of Gymnospermæ.

GNOMON'ICUS, (belonging to a dial) where any stalk-like appendage is abruptly bent at a high angle to the part to which it is attached.

GOBLET-SHAPED, hemisphærical, concave, and somewhat contracted below. Fig. 92.

92

GONGYLO'DES, (γογγυλης round, ειδος resemblance) knob-like.

GON'GYLUS, (γογγυλης round) rounded corpuscles produced on certain Algæ, which become ultimately detached, and germinate as distinct individuals. Globular bodies scattered through the thallus of Lichens. Synonyme for *SPORA*, *SPORIDEA*, and *SPEIREMA*.

GONNOPHO'RUM, (γονος generation, φέρω to bear) an elevated or elongated receptacle, bearing the stamens and carpels in a prominent and conspicuous manner.

GOODENIA'CEÆ, GOODENO'VIÆ, (from the genus Goodenia) a natural Order of Dicotyledones.

GOSSYP'INUS, *GOSSIP'INUS*, (made of cotton) cottony.

GOURD, a fleshy, one-celled, and many-seeded fruit, with parietal placentæ; the cell frequently filled with pulp when ripe. Ex. Melons, Gourds, Cucumbers. Fig. 93.

93

GRA'CILIS, slender.

GRAFT, the portion of one plant to be grafted on another plant, which is termed the Stock.

GRAMI'NACEÆ, GRAMI'NEÆ, (*GRAMEN* grass) the Grass Tribe. A very extensive Natural Order of Monocotyledones.

GRAM'MICUS, (made by lines) lettered.

GRANIF'ERUS, (*GRANIFER* bearing grains of corn) synonyme for Monocotyledones.

GRAN'ULAR, GRAN'ULATED, *GRANO'SUS*, *GRANULA'TUS*, *GRANULO'SUS*, (*GRA'NUM* a grain of corn, or kernel of a fruit) when any organ is covered with, or is composed of, small tubercles resembling grains.

GRAN'ULE, *GRAN'ULUM*, (diminutive of *GRANUM* a grain) a small grain, many of which are contained in each grain of pollen, and constitute the fovilla. A large kind of sporule found in some Algæ; also, a sporule of all cryptogamic plants. A small wort-like appendage, of which there are one or more on the calyx of certain species of Rumex.

GREA'SY, where the surface feels as though it were rubbed with grease.

GREY, *GRI'SEUS*, the neutral tint, which may be formed by mixing blue, red, and yellow, in equal proportions.

GROSSIFICA'TION, *GROSSIFICA'TIO*, the process of swelling in the ovary, after fertilization.

GROSSULA'CEÆ, GROSSULA'RIEÆ, (from the old genus Grossularia, now Ribes) the Currant Tribe. A natural Order of Dicotyledones.

GRUINA'LIS, (*GRUS* a crane) shaped like the bill of a crane; as the fruit of the Geranium Tribe.

Gru'mous, *Grumo'sus* (*Grumus* a hillock of earth) In clustered grains. Applied to clustered fleshy tubercular roots, as those of Ranunculus ficaria. Synonyme for Granular.

Gum, (*Gummi*) a vegetable secretion which may be detected in the sap of most plants, and which is excreted by many, and hardens on their surface.

Gutta'tus, spotted.

Guttif'eræ, (from Gutta a drop, because most of the species yield a Gum-resin allied to Gamboge) the Mangosteen Tribe. A natural Order of Dicotyledones.

Gymnocar'pus, (γυμνος naked, καρπος fruit) where the pericarp is either without any pubescence; or where it does not adhere to any of the outer floral whorls.

Gymnosper'mia, (γυμνος naked, σπερμα a seed) an order of the artificial class Didynamia; where the fruit is formed of four, more or less distinct, nut-like carpels, surrounded by the persistent calyx. Each carpel was formerly considered to be a separate naked seed.

Gymnosper'mous, *Gymnosper'mus*, (γυμνος naked, σπερμα a seed) where the ovules are developed without the usual integumentary accompaniment of a pericarp, as in the Coniferæ.

Gym'nosperms, *Gymnosper'mi*, considered either as a natural Order of Dicotyledones, or as a separate Class; where the ovules are gymnospermous.

Gynan'dria, (γυνη a woman, ανηρ a man) an artificial class of the Linnean system, where the stamens are so far united to the carpels that both together form a central column surrounded by the perianth.

Gynan'drous, *Gynan'dricus*, *Gynan'drus*, where the stamens and carpels cohere, as in Gynandria.

Gyni'zus, (γυνη a woman) synonyme for stigma; applied only in the family of Orchideæ.

Gynœ'cium, (γυνη a woman, οικος a house) the aggregate of the carpels, composing the innermost of the floral whorls; synonymous with Pistillum when the carpels cohere.

Gyno'basis (γυνη a woman, βασις a basis) the dilated base of a solitary style, surmounting a multilocular ovary.

Gyno'phore, *Gyno'phorus*, (γυνη a woman, φερω to bear) a pedicellary support to the ovary, seated on the receptacle.

Gynoste'gium, (γυνη a woman, στεγη a covering) synonyme for perianth.

GYNOSTE'MIUM, (γυνη a woman, στημων a stamen) the columnar mass formed by the union of the style and filaments in Orchideæ.

GYP'SEUS, (*GYPSUM* white lime) synonyme for Cretaceus.

GYRA'TUS, (turned about) synonyme for Circinatus.

GYRO'MA, (*GYRUS* a circle) synonyme for Annulus.

HABIT, (*HA'BITUS*) the peculiar aspect assumed by any species, depending upon the aggregate of its natural characters.

HABITA'TION, (*HABITA'TIO*) a term used in Botanical Geography to signify the limits within which a particular species is found naturally distributed on the earth's surface.

HÆMATI'TICUS, (αιματιχος bloody) brown-red; red with much grey.

HÆMATO'PHYLLUS, (αιμα blood, φυλλον a leaf) where a leaf is marked with red blotches.

HÆMODORA'CEÆ, (from the genus Hæmodorum) the Bloodroot tribe. A natural Order of Monocotyledones.

HÆMORRHA'GIA, (a continued flux of blood) a disease in plants where the sap is continually exuding through an external wound.

HAIR, expansions of cellular tissue, in the form of hairs, &c. which coat the surface of various parts of many plants.

HAIR'INESS, *HIRSU'TIES*, (*HIRSUTUS* hairy, rough) where the hair is less soft and longer than in the form termed "pubescence" or "down."

HAIR-POINTED, terminating in a fine and weak hair-like point.

HAIR-SHAPED, synonyme for Capillary.

HAIRY, having hair of the character expressed by "hairiness."

HALBERT-HEADED, synonyme for "hastate."

HALESIA'CEÆ, (from the genus Halesia) synonyme for Styraceæ.

HALF-NETTED, where the outermost only of several investing layers is reticulate.

HALF-TERETE, flat on one side, terete on the other; like one half of a cylinder which has been divided longitudinally through the axis.

HALONA'TUS, (ἁλος a button or stud) when a coloured circle surrounds a spot.

HALO'PHYTON (ἅλς the sea, φυτον a plant) a plant which grows within the influence of sea water.

HALORA'GEÆ, (from the genus Haloragis) a natural group of

Dicotyledones; considered either as a separate order, or as a Sub-Order of Onagraceæ.

HALVED, synonyme for Dimidiate.

HAMAMELA'CEÆ, HAMAMELI'DEÆ, (from the genus Hamamelis) the Witch-Hazel Tribe. A natural Order of Dicotyledones.

HAMA'TUS, *HAMO'SUS*, hooked.

HA'MULUS, *HA'MUS*, a hook.

HAPLOGE'NEUS, (ἁπλόος simple, γενναω to beget) synonyme for Heteronemius.

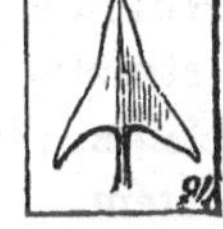

HAS'TATE, *HASTI'LIS*, *HASTA'TUS*, (bearing spears) shaped like the head of a halbert; the base diverging on each side into an acute lobe. Fig. 94.

HEADED, synonyme for "Capitate."

HEART-WOOD, the innermost layers of wood in Exogenous trees; which have become more deeply coloured than the outermost, and much harder.

HEART SHAPED, synonyme for "Cordate."

HEBECAR'PUS, (ἥϐη down, καρπος fruit) where a fruit is covered with a downy pubescence.

HEBETA'TUS, (made blunt) where the extremity is obtuse and more or less soft.

HEL'MET, an arched concave petal or sepal; or a portion of one; as the upper lip of several labiate flowers. Fig. 95, *h*.

HELVO'LUS, (pale red) yellowish-red and grey.

HEMEROCALLI'DEÆ, (from the genus Hemerocallis) a natural group of Monocotyledones, considered to be either a distinct order, or a section of the Order Liliaceæ.

HEMICAR'PUS, (ἥμισυς half, καρπος fruit) one portion of a fruit which spontaneously divides into two separate parts; as that of Umbelliferæ.

HEMICYLIN'DRICUS, (ημισυς half, κύλινδρος a cylinder) synonyme for "half-terete;" also, a foliaceous expansion, plane on one side and convex on the other.

HEMIGY'RUS, (ημισυς half, γῦρος a circle) a pericarp formed like a follicle, but more or less hard and woody; as in Proteaceæ.

HENSLOVIA'CEÆ, (from the genus Henslovia) a natural Order of Dicotyledones, containing only the genus Henslovia.

HEPA'TICÆ, (*HEPATICUS*, of the liver) the Liver-wort tribe. A natural Order of Acotyledones.

Hepa'ticus, (of the liver) liver-coloured. Yellowish-red with much grey.

Heptagy'nia, (ἑπτὰ seven, γυνὴ a woman) an artificial Order in the Linnean system, characterized by the presence of seven pistils, or at least by seven distinct styles.

Heptan'dria, (επτα seven, ανηρ a man) an artificial class in the Linnean system, characterized by an hermaphrodite flower possessing seven stamens. Also, an order in the classes of the same system where the flowers are unisexual.

Heptand'rous, *Heptan'der*, *Heptan'drus*, possessing the structure described under Heptandria.

Herb, (*Herba*) a plant which does not possess a woody stem.

Herba'ceous, (*Herbaceus* belonging to herbs) used in contradistinction to woody. Also, applied to any portions of a plant which are more particularly green and succulent.

Herba'rium, (*Herbarius* belonging to herbs) a collection of plants, properly dried and prepared for botanical study.

Hermaph'rodite, *Hermaphrodi'tus*, where both stamens and pistil occur in the same flower.

Hernandia'ceæ, Hernan'dieæ, (from the genus Hernandia) a natural Order of Dicotyledones.

Hesperideæ, (from a supposition that oranges grew in the garden of the Hesperides) synonyme for Aurantiaceæ.

Hesperi'dium, (fruit of Hesperideæ) an indehiscent many-celled fruit, coated with a spongy rind; the cells containing a mass of pulp, in the midst of which a few seeds are embedded.

Heterocar'pian, *Heterocar'pinus*, (ετερος another, καρπὸς fruit) an inferior, or rather, a partially inferior fruit, as in the Acorn.

Heterocar'pus, (ετερος another, καρπὸς fruit) where a plant bears fruit more or less distinguishable into two separate forms.

Heterocli'tus, (ετερος another, κλιτος a declivity) deviating from the ordinary rule. Where plants have monœcious or diœcious flowers.

Heterog'amus, (ετερος another, γαμος a wedding) bearing flowers of different sexes.

Heterone'meus, (ετερος another, νημα a thread) Acotyledonous plants whose sporidia germinate in the form of threads, which afterwards unite to form a heterogeneous body; as Ferns and Mosses.

HETER'OTROPAL, *HETER'OTROPUS*, (ετερος another, τρέπω to turn) where the embryo lies oblique or transverse to the axis of the seed, the radicle not being directed to the hilum. Fig. 96.

HEXAGY'NIA, (εξ six, γυνη a woman) an Order in the artificial system of Linneus, characterized by the flower having six pistils.

HEXAG'YNOUS, *HEXAG'YNUS*, (εξ six, γυνη a woman) having the structure explained under Hexagynia.

HEXAN'DRIA, (εξ six, ανηρ a man) an artificial class in the Linnean system, characterized by hermaphrodite flowers which have six stamens. Also, an Order in the same system, where the classes depend upon something more than the mere number of the stamens.

HEXAN'DROUS, *HEXAN'DRICUS*, *HEXAN'DRUS*, (εξ six, ανηρ a man) possessing six stamens as in Hexandria.

HEXAR'RINUS, (εξ six, ἄρρην a male) synonyme for Hexandrus.

HEXASTE'MONIS, (εξ six, στημων a stamen) synonyme for Hexandrous.

HIANS, Gaping.

HID'DEN, where some part is so completely invested by another that it is concealed from sight; as where the radicle of Tropæolum is concealed by the prolonged bases of the cotyledons.

HIDDEN-VEINED, synonyme for veinless; where veins really exist, but are concealed by an excess of parenchyma.

HILA'RIS, belonging to the Hilum.

HILOSPER'MEÆ, (from the large size of their hilum) a synonyme for Sapoteæ.

HI'LE, *HI'LUS*, (*HILUM* the black scar on the surface of a bean) the scar left on the surface of a seed at the spot where it was attached by the funiculus or umbilical chord to the pericarp.

HINOI'DEUS, (ινοειδης nerved) synonyme for "divergent curvinerved" leaves. See Fig. 64, *b*.

HIPPOCASTA'NEÆ, (from the species Æsculus hippocastanum) the Horse-chesnut tribe. A natural Order of Dicotyledones.

HIRCI'NUS, (of a goat), *HIRCO'SUS*, (smelling like a goat) a goat-like smell peculiar to several plants.

Hirsu'ties, hairiness.

Hirsu'tus, hairy.

Hirtel'lus, *Hirtus*, (hairy) shaggy.

His'pid, *Hispido'sus*, *Hispid'ulus*, (*His'pidus* rough) where the pubescence is composed of long and rigid hair.

Hoa'ry, greyish white; i. e. a light grey, an appearance originating from white pubescence.

Homalia'ceæ, Homali'neæ, (from the genus Homalium) a natural Order of Dicotyledones.

Homoge'neal, *Homo'genus*, (ὁμος like, γενος a race) of the same nature or kind.

Homoneme'æ, (ὁμὸς alike, νημα a thread) Acotyledonous plants, composed of filaments, which in germination separate from a homogeneous body—as in Algæ and Fungi.

Homope'talus, (ομὸς alike, πεταλον a petal) either, a plant where all the petals are formed alike; or, the receptacle of a composite flower where all the florets are alike.

Homo'tropal, Homo'tropous, *Homotro'pus*, (ομος alike, τρεπω to turn) when the embryo is not straight, but still has the same general direction as the seed. Fig. 97.

Hon'ey-comb'ed, synonyme for "Alvèolate."

Hood, a concave form of sepal or petal resembling a monk's hood, as in the Aconites. Also, a peculiar expansion of the filaments where they unite and cover the ovary, as in Asclepias syriaca.

Hood'ed, synonyme for Cucullate. Fig. 61.

Hook, a stiff hair, bristle, or prickle, curved back at the point.

Hook'ed, formed as in the hook.

Hook'ed-back, synonyme for runcinate.

Ho'rary, *Hora'rius*, (hourly) lasting about an hour—as the flowers of certain plants in their expanded state.

Hordea'ceus, (of barley) shaped like a spike of barley.

Horizon'tal, *Horizonta'lis*, when a plane surface lies perpendicularly to the axis of the body which supports it; as in most leaves. Or, when one part is perpendicular, or nearly so, to another, whose natural position is usually vertical; as in branches with respect to the main stem.

Horn, any appendage which is shaped somewhat like the horn of an animal; as the spur of the petals in Linaria.

Horn'ed, terminating in a horn.

Hor'nus, (of the year) any part of a plant developed during the year.

Hor'ny, of a hard close texture, resembling horn in its consistency.

Hor'tus-sic'cus, (*Hortus* a garden, *siccus* dry) synonyme for Herbarium.

Hosto'rium, (*hostio* to requite) the absorbing organ of parasites, which supplies the place of a root.

Hugonia'ceæ, (from the genus Hugonia) a natural Order of Dicotyledones, restricted to the single genus Hugonia.

Humifu'sus, (*Humus* the ground, *fusus* laid all along) Procumbent.

Hu'milis, Low.

Humiria'ceæ, (from the genus Humirium,) a natural Order of Dicotyledones.

Hyacin'thinst, *Hyacin'thus*, (a violet or purple flower) *Hyacin'thinus*, (of a violet or purple colour) blue with a violet tinge. Blue with a little red.

Hya'linus, (ὑαλινος glassy, transparent) more or less transparent or translucent.

Hyberna'culum, (*Hybernacula* winter quarters) any part which invests the nascent shoot, and protects it during the winter, as buds and bulbs.

Hy'brid, *Hy'bridus*, (*Hybrida* a mongrel) the common offspring of two distinct species.

Hydral'gæ, (ὑδωρ water, *alga* a sea-weed) synonyme for Hydrophyte.

Hydroce'reæ, (from the genus Hydrocera) synonyme for Balsaminaceæ.

Hydrochara'ceæ, Hydrochari'deæ, Hydrocha'rides, (from the genus Hydrocharis) the Frog-bit tribe. A natural Order of Monocotyledones.

Hydrolea'ceæ, (from the genus Hydrolea) a natural Order of Dicotyledones

Hydrophylla'ceæ, Hydrophyl'leæ, (from the genus Hydrophyllum) the Water-leaf tribe. A natural Order of Dicotyledones.

Hydro'phyte, (*Hydro'phyton*, ὑδωρ water, φυτον a plant) an aquatic Alga.

Hydropte'rides, (ὑδωρ water, πτερις a fern) synonyme for Marsilliaceæ.

HYGROBI'EÆ, ὑγρον water, βιοω to live) synonyme for Haloragеæ.

HYGROME'TRIC, *HYGROME'TRICUS* (ὑγροὺ water, μετρεω to measure) expressive of the state or degree of hygroscopicity of a plant.

HYGROSCOPI'CITY, *HYGROSCOPI'CITAS*, (ὑγρον water, ϛκοπεω to observe) the property by which vegetable tissues absorb or discharge moisture, according to circumstances.

HYME'NEUM, *HYME'NIUM*, (ὑμήν a membrane) that portion of the fructification of a Fungus in which the sporules are situated; usually more or less membranous.

HYMENO'MYCETES, *HYMENO'MYCI*, (ὑμην a membrane, μυκης a fungus) fungi furnished externally with a sporuliferous Hymenium.

HYPAN'THIUM, *HYPANTHÆ'DIUM*, (ὑπο under, ἄνθος a flower) A fleshy receptacle without any involucrum; as in the Fig.

HYPERBORE'AN, *HYPERBO'REUS*, (far northward) indigenous in the northernmost countries, within the Arctic circle.

HYPE'RICÆ, HYPERICA'CEÆ, HYPERICI'NEÆ, (from the genus Hypericum) the Tutsan tribe. A natural Order of Dicotyledones.

HY'PHA, (ὑφα a web) the filamentous, somewhat fleshy, deliquescent thallus of Byssaceæ, or plants which form mouldiness.

HYPHAS'MA, (ὑφασμα a web) the web-like thallus of agarics.

HYPHOMYCE'TES, (ὑφα a web, μυκης a mushroom) fungi whose thallus is of the form called Hyphasma.

HYPOBLAS'TUS (ὑπὸ under, βλαστος a shoot) synonyme for the fleshy cotyledon of the Graminaceæ.

HYPOCARPOGÆ'US, (ὑπὸ under, καρπος fruit, γη the earth) synonyme for Hypogæus.

HYPOCHI'LUS, (ὑπο under, χεῖλος the lip) the lower portion of the lip of Orchidaceæ.

HYPOCRATE'RIFORM, *HYPOCRATERIFOR'MIS*, (υπὸ under, κρατηρ a goblet, *FORMA* shape) a monopetalous corolla with the tube long and cylindrical, and limb flat, and spreading at right angles to it.

HYPOGÆ'AN, *HYPOGÆ'US*, (ὑπὸ under, γη the earth) plants which ripen their fruit under ground. Also, generally, of any part that grows under ground.

HYPO'GYNOUS, *HYPO'GYNUS*, *HYPOGY'NICUS*, (ὑπὸ under, γυνη a woman) seated below the base of the ovary, but not attached to the calyx; as the stamen and petal in Fig. 98.

HYPOPHYL'LIUM, *HYPOPHYL'LUM*, (ὑπὸ under, φυλλον a leaf.) a small sheath-like petiole seated below a peculiar leaf-like form of branch; as in Asparagus.

HYPOSTAM'INEUS, (ὑπὸ under, σταμην a stamen) a monopetalous flower with hypogynous stamens

HYPOTHE'CIUM, (ὑπὸ under, θηκη a case) the substance which immediately invests the perithecium of Lichens.

HYPOX'IDEÆ, (from the genus Hypoxis) a natural group of Monocotyledones, considered either as a distinct Order, or as a tribe of Amaryllidaceæ.

HYSTERAN'THOUS, *HYSTERANTHE'US*, (ὑστερος after, ἄνθος a flower) plants whose leaves expand after the flowers have opened.

HYSTRELLA. (ὑστερα the matrix) synonyme for Carpella.

IAN'THINUS, (ιανϑινος violet) synonyme for Violaceus.

ICOSANDRIA, (εικοσι twenty, ανηρ a man) the twelfth artificial class of the Linnean System; characterized by the flowers containing about twenty stamens or more, which are partially attached to the calyx (perigynous) and consequently seem to originate from it.

ICOSAN'DROUS, *ICOSAN'DER*, *ICOSAN'DRUS*, having the stamens as in Icosandria.

ICTE'RINUS, (*ICTERUS* a yellow bird) of a yellowish tinge.

IG'NEUS, (fiery) synonyme for Flammeus.

IGNIA'RIUS, (*IGNIA'RIUM* tinder) having the puff-like consistency of German tinder.

ILICI'NEÆ, (from the genus Ilex) synonyme for Aquifoliaceæ.

ILLECEBRA'CEÆ, ILLECE'BREÆ, (from the genus Illecebrum) the Knot-Grass Tribe. A natural Order of Dicotyledones.

ILLIGERA'CEÆ, ILLIGE'REÆ, (from the genus Illigera) a natural Order of Dicotyledones.

IMBER'BIS, (Beardless) without any beard.

IM'BRICATE, IMBRICA'TED, *IMBRICA'TUS*, (laid one over another like tiles on a roof,) *IMBRICATIVUS*, where many parts are regularly arranged, and one part partially overlaps another.

IMMARGINA'TE, *IMMARGINA'TUS*, (*IN* without, *MARGO* a border) where the edges of any organ are not characterized by any marked difference in structure from the middle portions.

IMMER'SED, *IMMER'SUS*, (immerged) growing entirely under water. Also, when one part or organ is completely embedded in another, as the sporidia of some lichens in the thallus.

IMMOVE'ABLE, (*IMMOBILIS*) where no particular motion can take place about the point of attachment; as, where anthers firmly adhere to the filament.

IM'PARI-PINNA'TUS, (*IMPAR* odd, *PINNATUS* winged) unequally pinnate.

IMPER'FECT, (*IMPERFECTUS*) where certain parts, usually present, are not developed. As the stamens in some, and the carpels in other flowers.

IM'PLEX, *IMPLICA'TUS*, (wrapped) synonyme for plicatus.

IMPLEX'US, entangled.

IMPREGNA'TION, *IMPREGNA'TIO*, synonyme for Fertilization.

IMPRES'SUS, (engraven, marked) marked with slight depressions.

INADHÆ'RENS, (*IN* not, *ADHÆREO* to adhere) free from all adhesion to contiguous parts.

INÆQUA'LIS, unequal.

INÆQUILA'TERUS, (*INÆQUALIS* unequal, *LATER* a side) synonyme for Inæqualis.

INA'NIS, (empty) when a stem has no pith, or only what is very soft and inconsiderable.

INANTHERA'TUS, (*IN* without, *ANTHERA* an anther) when the filament produces no anther.

INCA'NUS, hoary.

INCARNA'TUS, synonyme for Carneus.

INCH, rudely measured, at about the length of the first joint of the thumb.

INCI'SION, (*INCISIO*) an indentation along the margin of a thin or foliaceous organ.

INCI'SED, (*INCISUS* cut) synonyme for Cut.

INCLI'NED, INCLI'NING, (*INCLINANS*, *INCLINATUS*,) much bent out of the perpendicular, in a curved line, the convex side upwards.

INCLU'DED, (*INCLUSUS* enclosed) when one part does not extend or protrude beyond another by which it is surrounded; as when the stamens or style do not extend beyond the mouth of a monopetalous corolla.

INCOMPLE'TE, *INCOMPLE'TUS*, (*IN* not, *COMPLETUS* finished) where some part, usually present in allied species, is not developed in some particular case. Also, where an organ has the appearance of not having been fully developed: as where placentæ project into the cavity, but do not reach the axis of the pericarp, and which, consequently, is not completely divided into separate cells.

INCRASSA'TUS, (*IN*, and *CRASSUS*, thick) thickened.

INCREAS'ING, see "*ACCRESCENS*."

INCUM'BENT, (*INCUMBENS INCUBITUS*, leaning or resting upon.) Where the radicle is bent and pressed against the back of one of the cotyledons, in certain Cruciferæ, Fig. 99, *a*. The symbol (||O) expresses this. Applied to the anther, it implies the attachment to the filament to be at the back, and not at the base, Fig. 99, *b*. It is used also synonymously with "procumbent."

INCRUS'TED (*INCRUSTA'TUS* made into a hard crust.) Where an outer envelope is firmly attached to the part it covers; as when a pericarp invests the seed so closely that it seems to form a portion of it.

INCUR'VED, *INCURVA'TUS*, (*INCURVUS*, bent in.) Gradually bending from without inwards; as where the stamens curve towards the pistil.

INDEF'INITE, *INDEFIN'ITUS*, (*IN* not, *DEFINITUS* defined.) Where the number of any particular description of organ is either uncertain, or forms no positive character. Thus, the number of stamens in a flower beyond twelve is not used in their artificial classification. Applied to the Inflorescence it is employed synonymously with "centrifugal" or "indeterminate."

INDEHIS'CENT, *INDEHIS'CENS*, (*IN* not, *DEHISCO* to gape) without dehiscence, or regular line of suture.

INDETER'MINATE, *INDETERMINA'TUS*, (*IN* not, *DETERMINATUS* limited) synonyme for indefinite.

INDIG'ENOUS, *INDIGENUS*, (*INDIG'ENA* a native.) A vegetable which is the spontaneous production of any country.

INDIGO-COLOURED, (*INDIGOTICUS*.) A deep but dull blue. Blue with grey.

INDIRECTE'-VENOSUS. Where the lateral veins, in a leaf, run together at the extremities, and emit little veins.

INDIVID'UAL, *INDIVID'UUM*, (*INDIVIDUUS*, inseparable.) Whatever is capable of separately existing, and reproducing its kind. Thus, a seed or a leaf-bud are each nascent states of individual plants.

45

INDIVI'SUS, (undivided.) Entire.

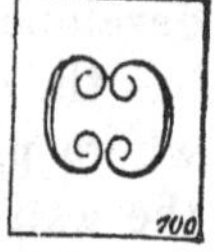

INDUPLICA'TE, *INDUPLICA'TUS*, (*IN* in, *DUPLICATUS* doubled.) Where the edges of a valvate estivation are folded inwardly, Fig. 100.

INDU'SIUM, (a shirt.) The cuticular covering to the sori of ferns.

INDU'VIÆ, (clothes.) Persistent portions of the perianth. Also the remains of certain leaves which not being articulated do not fall off and leave a scar.

INER'MIS. Unarmed.

INFE'RIOR, (*INFE'RUS.*) When one organ is placed below another. More especially used to express the connection of the tube of a calyx with the ovarium: when connected to the ovarium, so that it forms an outer coat to it whilst the limb stands on the summit, the calyx is called superior and the ovarium inferior, as in the Rose. When perfectly free without contracting any adhesion to the ovarium, the calyx is said to be inferior, and the ovarium superior, as in the Poppy. This term is applied to the radicle, when it is directed towards the hilum, at the base of the seeds.

INFLA'TUS, (swollen.) Bladdery.

INFLEX'ED, *INFLEX'US* (turned toward.) Synonyme for incurved.

INFLORESCENCE, *INFLORESCEN'TIA*, (*INFLORESCO* to flourish.) The general arrangement or disposition of the flowers in a plant.

INFRA-AXIL'LARY, *INFRA-AXILLA'RIS*, (*INFRA* below, *AXILLA* the armpit.) Originating below the axil of a leaf.

INFRAC'TUS, (much broken.) Synonyme for inflexus.

INFUNDIBULIFORM *INFUNDIBU'LIFORMIS*, (*INFUN'DIBULUM* a tunnel or funnel.) Shaped like a funnel, with a tube below and gradually enlarging upwards.

INNA'TE, *INNA'TUS*, (Inbred.) Where the point of attachment is at the apex; as in some cases of the anther to the filaments, Fig. 101.

INNOVA'TION, *INNOVA'TIO*, (*INNOVO* to make new.) An incomplete shoot; more especially the young shoots of mosses.

INOMY'CES (ἲς ἶνος a fibre, μυκης a mushroom.) An order of filamentous fungi.

INSERTION, *INSER'TIO*, (*INSERTUS* put in or upon.) The mode in which one body is connected to another, where it appears to have been attached to it, though in reality it has grown from it; as the leaf on the branch; the branch on the stem, &c.

In'teger. Entire.

Integer'rimus, (very entire.) Besides being perfectly free from incision, this term implies a slight thickening of the margin, as though it were bordered.

Integ'ument, (*Integumentum*.) A portion closely investing, or merely surrounding, another. Thus, the parts of the perianth are styled the "floral integuments" because they closely invest the stamens and pistils in the bud state. The seminal integuments are the coats which invest the kernel of the seed.

Intercel'lular, *Intercellula'ris*, (*Inter* between, *Cellula* a little cellar.) Between the cells of the cellular tissue. Thus, the little interstices left by the cells not accurately filling space, are termed "intercellular passages."

Interno'de, *Interno'dium*, (*Inter* between, *nodus*, a knot.) A part of the stem between two nodes.

Interruptedly-pinnate, *Interrupte-pinna'tus*. Where the pairs of leaflets in a pinnate leaf are alternately larger and smaller. See Guide to the Botanist, figure 95.

Interrupted, (*Interruptus*.) Where symmetry, or regularity of outline or composition is partially destroyed.

Intor'sio, (*Intorqueo* to turn or wind.) A twining.

Intrafolia'ceous, *Intrafolia'ceus*, (*Intra* within, *Folium* a leaf.) The position of some part with respect to the axilla of a leaf. Thus, when the stipules are united to the anterior portion of the petioles only, and are free above, they stand between the leaf and stem, as in Arenaria rubra.

Intrica'tus. Synonyme for Inplexus.

Introcur'vus, (*Intro* within, *curvus* bended.) Synonyme for Imflexus.

Intror'se, *Intror'sus*, (Inwardly.) Turned inwards, or towards the axis of the part to which it is attached.

Introve'nius, (*Intro* within, *vena* a vein.) Synonyme for "Avenius," where the veins of a leaf, though really present, are concealed by the thickness of the parenchyma.

Innunda'tus, (overflown.) Living completely submerged in water.

Inver'se, (*Inver'sus* inverted.) Where the apex of one organ is placed in an opposite direction to that of another with which it is contrasted. The "embryo" is said to be inverse when the radicle is directed away from the hilum towards a point immediately opposite to it, Fig. 102, (*h*) hilum, (*r*) radicle.

INVISIBLE. When some part is not yet sufficiently developed to be distinctly recognized.

INVOLUCEL'LUM, (*INVOLUCRUM* a cover.) A "partial" involucrum.

INVOLUCRA'TUS. Furnished with an involucre.

INVOLU'CRE, *INVOLUCRUM*, (a cover.) A whorl of bracts, free or united, seated on the peduncle; either near, or at some distance below the flower or flowers; synonyme for Indusium.

IN'VOLUTE, *INVOLU'TIVUS*, *INVOLU'TUS* (wrapt up.) Where the edges of a foliaceous organ are rolled inwards; as some leaves in vernation, some sepals, &c. in æstivation.

IRREG'ULAR, *IRREGULA'RIS*, (*IN* against, *REGULA*, a rule.) Exhibiting a want of symmetry, used also synonymously with unequal.

IRIDA'CEÆ, IRID'EÆ (from the genus Iris.) The Corn-flax tribe. A natural order of Monocotyledones.

IRRITABIL'ITY, *IRRITABIL'ITAS*, (*IRRITABILIS* irritable.) A vital property by which certain parts, in some plants, exhibit the phenomenon of spontaneous motion, when under the influence of particular stimuli. By some, this is considered to be merely an extreme case of "Excitability."

ISADEL'PHUS, (ἰσος equal, ἀδελφος a brother.) When the separate bundles of stamens in a diadelphous flower are equal or alike.

ISO'BRIOUS, *ISOBRIA'TUS* (ἰσος equal, βριάω to be strong.) Applied to the embryo of Dicotyledones; because both the dicotyledons seem to be developed with equal force.

ISOCH'ROUS, (ἰσος equal, χροα colour.) Possessing a uniformity of colour throughout.

ISODY'NAMOUS, *ISODY'NAMUS* (ἰσος equal, δυναμος power.) Synonyme for Isobrious.

ISOSTE'MONOUS, *ISOSTEMO'NUS*, (ἰσος equal, στημων a stamen.) A flower which has stamens equal in number to the petals.

ITHYPHYL'LUS, (ἰθυς straight, φυλλον a leaf.) Where a leaf is stiff and straight.

IVORY-WHITE. White, slightly tinged with yellow, and having a little lustre.

JASMINA'CEÆ, JASMI'NEÆ (from the genus Jasminum.) The Jasmine Tribe. A natural order of Dicotyledones.

JOINTED. Either applied to stems, and other parts, which appear to possess joints; or more properly to such as actually possess them.

JOINTS. Certain parts where the uniformity of the tissue is altered; and where it may readily be ruptured or fall asunder in decay.

JU'BA, (a mane; the tops of trees.) A loose panicle, as in the case of the male flowers of Zea Mays. Also a dense cluster of awns, as in the spikes of certain grasses.

JU'GUM, (a yoke.) A pair of the opposite leaflets in a pinnated leaf.

JU'LUS, (a catkin.) Synonyme for Amentum.

JUNCAGINA'CEÆ, (from certain analogies with Juncaceæ.) A natural order of Monocotyledones.

JUNCA'CEÆ, JUN'CEÆ, (from the genus Juncus.) The Rush Tribe A natural order of Monocotyledones.

KEEL. A projecting ridge, rising along the middle of a flat or curved surface. Also the two lowermost, and more or less combined, petals of a papilionaceous corolla.

KEELED. Furnished with a keel.

KERMESI'NUS. Synonyme for puniceus.

KERNEL. The embryo, with or without a perisperm, enclosed in the seminal integuments. In a lax sense, employed to signify any seed enclosed in a hard case; and it even includes some dry pericarps.

KNEE-JOINTED. Synonyme for geniculate.

KNEEPAN-SHAPED. Concavo-convex, and very thick

KNOT. A swelling in some stems where the attachment of the leaves takes place.

KNOTTED. Where a cylindrical body is swollen at intervals into knobs, somewhat resembling a knotted chord.

LABEL'LUM. Lip; or rather the lower lip only.

LABIA'TÆ (from labium, a lip; the corolla being bilabiate.) The mint tribe. A natural order of Dicotyledones.

LABIA'TE, *LABIA'TUS*, (*LABIOSUS*, full-lip Where a tubular calyx or corolla has the limb divided into two unequal portions, or lobes, which are placed, above and below, so as to imitate the lips of a mouth. Fig. 103.

103

LABIO'SE, *LABIO'SUS*, (full-lipped.) Where the petals of a

polypetalous corolla are so arranged as to imitate the Labiate form.

LA'BIUM. Lip.

LACERA'TUS. LA'CERUS. Torn.

LA'CHRYMÆFOR'MIS, (*LACHRYMA*, a tear, *FORMA*, shape.) Tear-shaped.

LACIN'IATE, *LACINIA'TUS*, (*LACINIA*, a fringe.) Fringed. Also "Slashed."

LACIN'ULA, (diminutive from *LACINIA*, a fringe.) The small inflexed point of the petals in Umbelliferæ.

LACISTEMA'CEÆ, LACISTE'MEÆ, (from the genus Lacistema.) A natural order of Dicotyledones.

LACTES'CENS. Producing milk.

LAC'TEUS, (like milk.) Milk-white.

LACU'NA, (a little hole.) Air-cell. Also small depressions on the upper surface of the thallus of Lichens.

LA'CUNOSE, *LACUNO'SUS*, (full of holes.) Where the surface is covered with depressions, (*LACUNÆ.*)

LACUS'TRINE, *LACUS'TRIS*, (*LACUS* a lake.) Living in or on the margins of lakes.

LÆVIGA'TUS. Polished.

LÆ'VIS. Smooth.

LAMEL'LA, (a thin plate of metal.) The Gill, in Agarics. Synonyme for Corona in some Silenaceæ.

LAMEL'LAR, *LAMELLA'TUS*, *LAMELLO'SUS.* (*LAMELLA* a thin plate of metal.) Tipped with two flat lobes, as are many styles.

LAMIA'CEÆ, (from the genus Lamium.) Synonyme for Labiatæ.

LAM'INA, (a thin plate of metal.) The Limb.

LANA'TUS. Woolly.

LAN'CEOLATE, *LANCEOLA'RIS*, *LANCEOLA'TUS*, (*LANCEA* a lance.) Shaped like the head of a spear; narrow and tapering at each end.

LANUGINO'SUS. (Downy.) Cottony.

LANU'GO, (tender hair.) Fine soft pubescence.

LAPI'DEUS, *LAPILLO'SUS*, (Stony.) Of a hard texture, like the nuts of stone-fruits.

LAPPA'CEUS, (like a bur.) Synonyme for Hamatus.

LARDIZABA'LEÆ, (from the genus Lardizabala.) A group of Dicotyledones, considered to be either a distinct order, or a section of Menispermaceæ.

LARVA'TUS (*LARVA* a mask.) Synonyme for Personatus.

LASIAN'THUS (λασιος hairy, rough; ἄνθος a flower.) When the pubescence on the flower is velvetty.

LATEBRO'SUS, (full of dens.) Hidden.

LA'TENT, *LA'TENS*, (lurking.) Lying dormant till excited by some particular stimulus; as the adventitious buds occasionally developed in trees.

LATERAL, *LATERA'LIS*, (of the side.) Fixed on, or near the side of any organ.

LATERINER'VIUS, (*LATUS* the side, *NERVUS* a nerve.) Synonyme for Rectinervius.

LATERI'TIUS, (made of brick.) Of a brick-red colour. Vermillion with much grey.

LA'TEX, (juice.) The proper-juice, or returning sap of plants. Also applied to the moisture which exudes from the stigma. Also the gelatinous matter surrounding the sporules of certain fungi.

LATISEP'TUS, (*LATUS* broad, *SEPTUM* a hedge.) Where the dissepiment in the fruit of Cruciferæ is broad in proportion to the thickness between the valves.

LAURA'CEÆ, LAURI'NEÆ, (from the genus Laurus.) The Cinnamon tribe. A natural order of Dicotyledones.

LAXUS, Loose.

LEAF. An appendage to the stem, considered as an expansion of the bark, composed of cellular tissue, and generally with fibres of vascular tissue intermixed.

LEAF-BUD. See Bud.

LEAFLET. Each separate portion or subordinate expansion in the limb of a compound leaf.

LEAF-LIKE. Synonyme for Foliaceous.

LEATHERY. With a consistency more or less resembling the toughness of leather.

LE'CUS, (λέχος a bed.) Synonyme for Cormus.

LECYDITHA'CEÆ, LECYTHIDEÆ, (from the genus Lecythis.) Either a section of Myrtaceæ, or a distinct order of Dicotyledones.

LEG'UME, *LEG'UMEN* (pulse.) The seed vessel of Leguminosæ. One celled and two-valved, with the seeds arranged along the inner angle; subject, however, to several modifications, which considerably mask the normal character.

LEGUMINO'SÆ, (*LEGUMEN* any kind of pulse.) The Bean or Pea tribe. An extensive order of Dicotyledones.

LEGU'MINOUS, *LEGUMINA'RIS*, (as in *LEGUMEN*.) When the dehiscence of a pericarp is by a marginal suture.

LEMNA'CEÆ (from the genus Lemna.) Synonyme for Pistiaceæ.

LENS-SHAPED. *LENTICULA'RIS*, *LENTIFOR'MIS*, (*LENS* a lentil, *FORMA* shape.) Of the form of a double-convex lens.

LENTIBULA'CEÆ, LENTIBULAR'IÆ, (from the lenticular shape of the air bladders on the branches of Utricularia.) A natural order of Dicotyledones.

LENTICEL'LA, (diminutive of *LENS*.) Small lens-shaped spots on the bark of many plants, from whence roots issue, under circumstances favorable to their development.

LEN'TICULAR. Synonyme for Lens-shaped.

LENTIGINO'SUS, (covered with freckles.) Dusty.

LEPAL, *LE'PALUM*. A nectary, originating in a barren transformed stamen.

LEPI'CENA, (λέπις a scale, κὲνος empty.) Synonyme for Gluma, as restricted to the outermost scales of the floret of Grasses.

LEPIDO'TUS, (λεπις a scale.) Synonyme for Leprosus.

LE'PIS, (dross scales in metal.) A scale.

LEPIS'MA, (λεπισμα peeled bark.) A cup-like form of disk surrounding an ovary.

LEP'ROUS, *LEPRO'SUS*, (*LEPRA* the leprosy.) Covered with the form of scale termed Lepis.

LETTERED. When superficial markings have an appearance of rudely formed letters.

LEUCAN'THUS, (λευκος white, ανθος a flower.) Bearing white flowers.

LI'BER. Bark.

LI'BER, *LIBERA'TUS*, (Free.) Separate.

LICHENA'CEÆ, LICHE'NÆ, LICHE'NES, (*LICHEN*, a liverwort.) The Lichen Tribe. An extensive order of Acotyledones.

LIG'NEOUS, *LIG'NEUS*, (Wooden.) Synonyme for Woody.

LIGNIF'EROUS, (*LIGNUM* wood, *FERO* to bear.) When branches form wood only, without flowers or fruit.

LIG'NINE, *LIG'NINA*, (*LIGNUM* wood.) A substance which fills the cellular tissue composing woody fibre.

LIGNO'SUS. Woody.

LIG'ULA (a shoe-strap.) A membranous appendage at the summit of the sheathing petiole in Gramineæ. Also, an appendage at the base of some forms of Corona.

LIGU'LATE, *LIGULA'TUS*, (*LIGULA*, a strap,) Synonyme for Strap-shaped.

LILAC, *LILACI'NUS* Blue and red, with a little grey.

LILIACEÆ, (from the genus Lilium). The Lily Tribe. A natural order of Monocotyledones.

LILIA'CEOUS, *LILIA'CEUS*, (*LILIUM* a lily.) The perianth ormed as in Liliaceæ.

LIMB, *LIMBUS* (a border). The superior portion (generally spreading) of a foliaceous organ, either sessile or surmounting a separate portion, by which it is connected with the axis. In the leaf, the latter portion is termed the petiole; in a petal, the claw, &c.

LIMBA'TUS (*LIMBUS* a border). Bordered.

LIMNANTHA'CEÆ, LIMNAN'THEÆ (from the genus Limnanthus). A natural order of Dicotyledones.

LINA'CEÆ, LI'NEÆ (from the genus Linum). The Flax-Tribe A natural order of Dicotyledones.

LINE, *LINEA* (a line). The twelfth part of an inch.

LINEA'LIS (*LINEA* a line). The length of a line.

LI'NEAR *LINEA'RIS* (of a line). Where the side margins of a foliaceous expansion are parallel, and the length considerably longer than the breadth. See Diagram, in the Guide, page 69.

LINED, *LINEA'TUS* (traced out). Synonyme for "striated."

LINEOLA'TUS (diminutive of *LINEATUS*). Marked with little lines.

LINGUIFOR'MIS, *LINGULA'TUS* (*LINGUA* the tongue, *FORMA* shape). Tongue-shaped.

LIP. Each of the two large lobes of a bilabiate perianth; as in the Labiatæ. Also applied to one of the segments of an irregular perianth; when this usually assumes some shape remarkably different from that of the other segments, as in Orchidaceæ.

LIPPED. Synonyme for "labiate."

LIREL'LA (diminutive for *LIRA* a ridge of land). Where the apothecia of a lichen are linear; as in Opegrapha.

LITHO'PHILUS (λιθος a stone, ιϑιλοςα a friend). Applied to plants which grow on bare rocks and stones.

LITTLE. Where the whole is small, but the several parts retaining the usual proportions.

LIT'ORAL, *LITORA'LIS* (of the sea-shore). Growing on the shores of the sea, or banks of rivers.

Litura'tus (*Litura* a blot). Superficial spots or blurs, as though the skin were abraded.

Liv'id, *Liv'idus*. Of a pale lead colour. Grey with blue.

Loasa'ceæ, Loa'seæ (from the genus Loasa). A natural order of Dicotyledones.

Lobe, *Lobus* (λοβὸς the tip of the ear). A rounded projecting part of some organ.

Lo'bed, *Loba'tus*. Divided into lobes, See Diagram, at page 80, of the Guide.

Lobelia'ceæ (from the genus Lobelia), A natural order of Dicotyledones.

Locu'lament, *Loculamen'tum* (a partition or apartment). A cavity in the pericarp containing the seed. One of the cells of the anther.

Locula'ris, *Locula'tus* (having distinct holes). Containing more than one loculament.

Loculici'dal, *Loculici'dus* (*Loculus* a cell, *cieo* to move). When the dehiscence takes place in the middle of the back of each loculament, along the dorsal nerve.

Loculo'sus (full of holes). Partitioned.

Lo'culus (a partition, a bag). A Loculament.

Loc'usta. Synonyme for "spicula." Also for "gluma."

Lodi'cula. Synonyme for Glumella.

Lofty. Very "Tall."

Logania'ceæ, Loga'nieæ (from the genus Logania). A natural order of Dicotyledones.

Lomenta'ceous, *Lomenta'ceus* (see *Lomentum*). When an expansion appears pinched at intervals, as though it were made up of several separate pieces applied end to end.

Lomen'tum (Bean-Meal). A "legumen" which is contracted in the spaces between the seeds.

Longitu'dinal, *Longitudina'lis* (*Longitudo* length). With reference to the axis of any part,

Longus, *Longissimus*. When some part is of greater length than some other part with which it is connected.

Loose. Of a soft texture, as though the separate parts were scarcely cohering. Also, when separate parts are arranged at some distance from each other upon a common axis.

Lorantha'ceæ, Loranthеæ (from the genus Loranthus), A natural order of Dicotyledones.

Lora'tus (*Lorum* a thong). Synonyme for "Ligulatus."

Lo'rica (a coat of mail). Synonyme for Testa.

Lo'rulum (diminutive for *Lorum* a thong). In lichens, when the thallus is filamentous and branching.

Low. When a plant is of smaller dimensions than other species with which it is most nearly allied.

Low'ered. Where the lower lip of a bilabiate corolla is inclined at about a right angle, or more, to the tube.

Lu'cens, *Lu'cidus* (bright). Shining.

Luna'te, *Luna'tus* (like a half-moon). Synonyme for Crescent-shaped.

Lu'nulate, *Lunula'tus*. Synonyme for Lunate.

Lu'rid, *Lu'ridus* (pale and dismal). Of a dingy brown. Grey with orange.

Lute'olus, *Lutes'cens* (Yellowish). A pale tint of yellow.

Lute'us. Yellow.

Luxu'riant, *Luxu'rians* (superfluous). Generally applied where a superabundance of nutriment causes the organs of nutrition to be more developed than those of fructification.

Lycopodia'ceæ, Lycopodi'neæ (from the genus Lycopodium). The Club-moss tribe. A natural order of Acotyledones.

Lygodysodea'ceæ (from the genus Lygodysodea). A natural order of Dicotyledones, restricted to a single genus; otherwise considered as a section of Rubiaceæ.

Lymph, *Lym'pha* (water). The ascending newly introduced Sap.

Lymph'æduct, Lympha'tic. Synonyme for Duct.

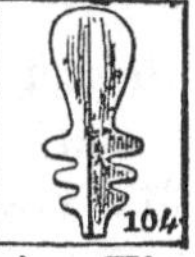

Lyra'te, *Lyratifi'dus*, *Lyra'tus* (*Lyra* a harp). Where a leaf has several pair of small lobes near the base, with deep sinuses between them, fig. 104.

Lythra'ceæ, Lythra'rieæ (from the genus Lythrum). The Loose-strife Tribe. A natural order of Dicotyledones.

Mace. The Arillus of the Nutmeg.

Macroceph'alous, *Macrocepha'lus* (μακρὸς large, κἑαλὴ a head). Where the cotyledons of a dicotyledonous embryo are confluent, and form a large mass compared with the rest of the body.

Macro'podous, *Macro'podus* (μακρὸς large, πους a foot). Where the radicle of a monocotyledonous embryo is large in proportion to the rest of the body.

Macula'tus, *Maculo'sus* (Spotted). Blotched.

Maculæfor'mis, *Maculifor'mis* (*Macula* a spot, *forma* shape). Where any part has the appearance of a mere spot; as the fructification of certain Algæ.

Magnolia'ceæ, Magno'liæ (from the genus Magnolia). The Magnolia Tribe. A natural order of Dicotyledones.

Male. A plant, or a flower, which bears stamens and no pistil.

Malesherbia'ceæ (from the genus Malesherbia). A natural order of Dicotyledones. Otherwise referred to a section of Passifloræ.

Malle'olus (the small shoots of a vine). The "layer" by which gardeners propagate plants.

Mallococ'cus (μαλλος a fleece, κοκκος a seed). Where the fruit is downy.

Malpighia'ceæ (from the genus Malpighia). The Barbadoes-Cherry Tribe. A natural order of Dicotyledones.

Malpighia'ceous, *Malpighia'ceus*. When hairs are formed as in the genus Malpighia; viz. attached by the middle, and lying parallel to the surface on which they grow.

Malva'ceæ (from the genus Malva). The Mallow Tribe. A natural order of Dicotyledones.

Mamil'la (a little teat). Little granular prominences on the surface of certain pollen.

Mamilla'ted, *Mamilla'ris*, *Mamilla'tus* (*Mamilla* a little teat). Where a wart-like projection surmounts a hemispherical body.

Mammo'sus (with large breasts). Synonyme for Mamillatus.

Mani'cate, *Manica'tus* (with sleeves or gloves). When pubescence is so much matted and interwoven that it may be easily removed from a surface in one mass.

Many-headed. When many distinct buds are seated on the crown of a root.

Maranta'ceæ, Maran'teæ (from the genus Maranta). The Arrow-root Tribe. A natural order of Monocotyledones.

Marbled. Stained with irregular streaks or veins of colour.

Marces'cent, *Marces'cens* (decaying). Gradually withering without falling off.

Marcgraavia'ceæ, Marcgravia'ceæ (from the genus Marcgraavia). A natural order of Dicotyledones.

Marchantia'ceæ (from the genus Marchantia). Synonyme for Hepaticæ.

Mar'gella (diminutive of Margo). The elliptic ring surrounding certain stomata.

Mar'ginal, *Margina'lis* (*Margo* an edge). Placed upon, or attached to, the edge of any thing.

Mar'gin (*Margo*). The boundary line or contour of a body, traced by the union of opposite plane surfaces.

Margina'rius (*Margo* an edge). Resulting from some modification of the marginal parts.

Margina'tus (Broad-brimmed). Edged.

Mari'ne, *Mari'nus* (inhabiting the sea), **Mar'itime**, (*Marit'imus*, belonging to the sea). Growing within the immediate influence of the sea. The former term is more frequently restricted to submerged plants, the latter to such as grow on the shore; but they are often used indiscriminately.

Marmora'tus, (covered with marble). Marbled.

Marsilia'ceæ (from the genus Marsilea.) The Pepperwort tribe. A natural order of Acotyledones.

Mas. Male.

Masculi'nus (Masculine). Possessing perfect anthers.

Mask'ed. Synonyme for Personate.

Mas'sula (a little lump). One of the smaller fragments which together compose the pollen mass in Orchidaceæ.

Mast. The acorn of the Beech.

Matura'tion, *Matura'tio* (a hastening). The process of ripening. Also the time when fruits are ripe.

Matuti'nus (of the morning). Taking place in the morning only, as the expansion of certain flowers.

Mea'ly. Covered with a scurfy powder. Possessing the texture and general appearance of flour.

Mea'tus-Intercculla'ris. Intercellular-passage.

Media'nus (middle). When some part originates or is connected with the middle of some other.

Mediterra'neus (*Medius* the middle, *terra* the earth). Inhabiting spots at a distance from the sea. Applied also to plants found exclusively in the neighbourhood of the Mediterranean Sea; or even belonging to the Flora of that district.

Medul'la. Pith. Also, in the seed, employed as a synonyme for Perispermium.

Medul'lary, *Medullo'sus* (full of marrow). Synonyme for Pithy.

Medul'lary-rays. Vertical plates of cellular tissue, which proceed from the pith to the surface, and are characteristic of the stems of Exogenæ.

Medul'lary-sheath. A thin zone of vascular tissue immediately surrounding the pith in Exogenous stems.

Medul'lina. Synonyme for Medulla.

Megaceph'alus (μεγας great, κεφαλη a head). Where the capitula of Compositæ, or heads of other flowers, are large.

Meioste'monous, *Meioste'monus* (μειων less, στημων a stamen). Where the stamens are fewer in number than the petals.

Melanophyl'lus (μελας black, φυλλον a leaf). Having leaves of a dark colour.

Melantha'ceæ, Melan'theæ (from the genus Melanthium). The Colchicum tribe. A natural order of Monocotyledones.

Melastoma'ceæ, Melasto'meæ (from the genus Melastoma). A natural order of Dicotyledones.

Melia'ceæ, Me'liæ (from the genus Melia). The Bead-tree tribe. A natural order of Dicotyledones.

Mel'inus (*Mel* honey). Of a honey colour.

Meloni'dium (μῆλον an apple, εἶδος form). Synonyme for Pomum.

Mel'on-sha'ped, *Melonifor'mis* (*Melo* a melon, *Forma* shape). Sphæroidal and longitudinally ribbed.

Membrana'ceous, *Membrana'ceus* (like parchment). Thin, and more or less transparent.

Mem'brane, *Membra'na*. A delicate pellicle of homogeneous tissue. Also a very thin layer composed of cellular tissue.

Membra'nula (diminutive from Membrana a Membrane). Synonyme for Indusium.

Memecyla'ceæ, Memecy'leæ (from the genus Memecylon). A natural order of Dicotyledones.

Meniscoi'd, *Meniscoi'deus* (μηνισκος a crescent, εἶδος resemblance). Shaped like a meniscus, or concavo-convex lens.

Menisperma'ceæ, Menisper'meæ (from the genus Menispermum). The Cocculus tribe. A natural order of Dicotyledones.

Menstrua'lis, Menstruus (monthly). Existing for about a month.

Mentzelia'ceæ (from the genus Mentzelia). A synonyme for Loasaceæ.

Merende'reæ (from the genus Merendera). A synonyme for Melanthaceæ.

Merecar'pium (μερις a portion, καρπος fruit). One carpel, with part of the calyx investing it, in the fruit of Umbelliferæ.

MERIDIA'NUS (noon-tide). Towards the South.

MERITHAL'LUS (μερις a portion, θαλλος a bough). Synonyme for Internodium.

ME'SOCARP, *MESOCAR'PUM* (μεσος middle, καρπος fruit). Synonyme for Sarcocarp.

MESOPHYL'LUM (μεσος middle, φυλλον a leaf). The whole inner portion or parenchyma of leaves, situate between the upper and under epidermis.

MESOSPER'MUM (μεσος middle, σπερμα seed). Synonyme for Sarcodermis.

METEOR'IC (*METEOR'ICUS*). Applied to flowers whose expansion is influenced by the state of the weather.

MICRO'BASIS (μικρος small, βασις a base). A variety of the Carcerulus, found in Labiatæ, where the gynobasis is very small, and surmounts a quadrilocular ovary; the cells of which are indistinct in the early stages.

MI'CROPYLE, *MICRO'PYLA* (μικρος small, πυλη a gate). The nearly closed foramen, as it exists in the ripened seed.

MID'RIB. The principal nerve or vein; which runs from the base to the apex of a leaf.

MILIA'RIS, (*MILIUM* a Millet seed). Applied to minute glandular spots on the epidermis, which seem to be generally the same as stomata.

MILK. An opaque white proper juice, found in many plants.

MILK-WHITE. Dull white with a feint bluish tinge.

MILL-SAIL-SHAPED. Applied to certain fruits (as of some Umbelliferæ), which have membranous expansions (or wings) disposed longitudinally upon their surface, fig. 105.

105

MINIA'TUS (of a vermillion colour). Pure red with a little yellow.

MINU'TE, *MINU'TUS* (lessened). Small compared with something analagous.

MIROBOLA'NEÆ (from Mirobolanus, a synonyme for the genus Terminalia). Synonyme for Combretaceæ.

MI'TRA (a bonnet). Used synonimously with galea, for "Helmet." Also, the thick, rounded, and folded pileus of some fungi.

MI'TRIFORM, *MITRÆFOR'MIS* (*MITRA* a bonnet, *FORMA* shape). Conical, and somewhat dilated at the base.

MIXTINER'VIS, *MIXTINER'VIUS* (*MIXTUS* mixed, *NERVUS* a nerve). A synonyme for "*RETICULATUS*" and "*VENOSUS*" when applied to the more common nervation of Dicotyledonous leaves.

Mo'bilis (moveable, variable). Synonyme for *versatilis*. Also applied to changeable (or shot) colouring. Also when some part has become partially detached, but is not entirely removed; as when the annulus of an agaric is detached from the stipes, but its remains are still about it.

Modiolifor'mis (*Modiolus*, the nave of a wheel). Nave-shaped.

Molendina'ceus, Molendina'ris (*Molendinarius* belonging to a mill). Mill-sail-shaped.

Mollis: Soft.

Monadel'phia (μονος alone, αδελφος a brother). An artificial class of the Linnean System, characterized by the stamens having their filaments united together, so as to form a ring or cylinder round the pistil.

Monadel'phous, *Monadel'phicus, Monadel'phus*. Where the stamens are all united, as in the class Monadelphia.

Monan'dria (μονος alone, ανηρ a man). An artificial class in the Linnean system, containing plants whose flowers have only one stamen.

Monan'drous, *Monan'der, Monan'dricus, Monan'drus*. A flower which contains only one stamen.

Monan'thus (μονος alone, ανθος a flower). Either where each peduncle bears a single flower; or, where the plant produces only one flower.

Monil'iform, *Monilifor'mis* (*Monile* a necklace, *forma* shape). Where a cylindrical body is contracted at regular intervals.

Monimia'ceæ, Monim'ieæ (from the genus Monimia). A natural order of Dicotyledones.

Mono'basis (μονος alone, βασις a base). Where the root is reduced to a small unbranched portion, as though it formed merely the base of the stem.

Monocar'pean, Monocar'pous, *Monocar'peus* (μονος alone, καρπος fruit). When a plant bears fruit once only during its existence; as annuals and biennials.

Monoceph'alous, *Monoceph'alus* (μονος alone, κεφαλη a head). Where flowers are disposed in single heads or umbels, &c. Also, where the ovary is surmounted by a solitary style or stigma.

Monochlamy'deous, *Monochlamy'deus*(μονος alone, χλαμυς

cloak. Where a flower has one whorl only to the perianth.

Monocli'nus (μονος alone, κλινη a bed). Synonyme for Hermaphrodite. Also applied to those capitula of Compositæ which consist of hermaphrodite florets only.

Monocotyle'dons, *Monocotyle'dones*. Plants having only one cotyledon.

Monody'namus (μονος alone, δυναμος power). Where one stamen is naturally much longer than the rest.

Monœ'cia (μονος alone, οικος a house). An artificial class, (as well as an order) in the system of Linneus; containing plants with two kinds of unisexual flowers on the same individual.

Monœ'cious, *Monoi'cus*. Possessing the characters explained under Monœcia.

Monoepigy'nia, (μονος alone, ἐπὶ upon, γυνὴ a woman). A class in the system of Jussieu, containing monocotyledonous plants whose stamens are epigynous.

Monoga'mia (μονος alone, γαμος marriage). An artificial order of the Linnean system, referred to the class syngenesia, and composed of certain plants whose flowers are not aggregated into heads, but whose anthers are more or less adhering. The order is no longer admitted, and these plants are referred to other classes.

Monog'enus (μονος alone, γενος a race). Synonyme for monocotyledonous; because such a plant has only one increasing surface in the centre.

Monogyn'ia (μονος alone, γυνη a woman). An artificial order in some of the classes of the Linnean system, characterized by the flowers having only one style or one pistil.

Monog'ynous, *Monogyn'icus*, *Monog'ynus*. With the characters of monogynia.

Monohypogyn'ia (μονος alone, ὑπὸ under, γυνη a woman). A class in the system of Jussieu, containing monocotyledonous plants, whose stamens are hypogynous.

Monoi'cus. See Monœcious.

Monoloc'ular *Monolocula'ris*. (μονος alone, *locula* a cell). Synonyme for Unilocular.

Monopet'alous, *Monopet'alus* (μονος alone, πὲταλον a petal.) Usually applied in the sense of "gamopetalous;" but also may be referred to the few examples of flowers whose corolla consists of a single laterally placed petal.

MONOPHYL'LOUS, *MONOPHYL'LUS*, (μονος alone, φυλλον a leaf). Synonyme for Monosepalous. Applied also to an involucrum composed of a single piece; or to a leaf bud, where a single leaf is subtended by an investing stipule.

MONOSEP'ALOUS, *MONOSEP'ALUS*, (μονος alone, σεπαλον a sepal). When the sepals, or subordinate parts of the calyx, are more or less united into a tube; in this case the outer perianth is "gamosepalous". But the term may also be applied to the rare case of calyces which have only one laterally placed sepal.

MONOSPER'MICUS, *MONOSPER'MUS* (μονος alone, σπερμος a seed). A fruit which contains only one seed.

MONOTROPA'CEÆ, MONOTROPEÆ (from the genus Monotropa). A natural order of Dicotyledones.

MON'STER, (*MON'STRUM*). When the conformation of any plant, or part of a plant, deviates from the usual and natural structure, the result is termed a monstrosity, (*MONSTRUOS'ITAS*); and the plant itself is called a "Monster."

MORINGA'CEÆ, MORIN'GEÆ (from the genus Moringa). A natural order of Dicotyledones, restricted to a single genus.

MORPHOLOGY, *MORPHOLO'GIA* (μορφη form, λογος a discourse). That department of botany which treats of the modification of certain fundamental organs; by which these are enabled to assume other functions than are performed by them under their normal condition.

MOSCHA'TUS (*MOSCHUS* musk). Possessing the odour of Musk.

MOUN'TAINOUS. See Alpine.

MOVE'ABLE (*MO'BILIS*). Applied to the anther, indicates its attachment to the filament to be so slight, that it is able to be moved in opposite directions. Applied to colours it refers to the shot or changeable character which these sometimes exhibit. The ring of an agaric is "moveable" when it detaches itself from the stipes and remains free about it.

MUCH-BRANCHED. When branches are very numerous and subdivided.

MU'COUS, *MUCO'SUS* (slimy). Of the nature of Gum. Also any thing coated with some secretion having this character.

MU'CRO (a sharp point). A straight stiff and sharp point.

MU'CRONATE, *MUCRONA'TUS* (pointed). Abruptly pointed by a sharp spinous process, (see Guide, p. 91, fig. 234.)

MUCRONULA'TIS (diminutive of *MUCRONATIS*). Where the plant is small.

MULTAN'GULAR, *MULTANGULA'RIS* (*MULTANGULUS*). With many angles; either in a solid body, as the stem, or a plain one, as a flat leaf.

MUL'TICEPS. Many-headed.

MULTI-DIGITATO-PINNA'TUS. When there are many secondary petioles, with a " Digitato-pinnate" arrangement.

MULTIF'ERUS (*MULTUS* many, *FERO* to bear). Often-bearing.

MULTIF'ID, *MULTIF'IDUS* (many-cleft). Where subdivisions or laciniations are deep and numerous.

MULTIJUGA'TUS, *MULTIJU'GUS*, (*MULTUS* many *JUGUM* a yoke). When a pinnate leaf bears many pairs of leaflets.

MULTILO'CULAR, *MULTILOCULA'RIS* (*MULTUS* many, *LOCULA* a small place; a cell). A cavity, subdivided by partitions into several cells: as in many seed vessels.

MULTIPARTI'TUS (*MULTUS* many, *PARTITUS* divided). Deeply divided into several strips or portions.

MUL'TIPLEX (many times as much). Where many of the same parts or organs occur together.

MULTIPLICA'TUS (multiplied). Where the petals of double flowers arise from supernumerary developements of the parts of the floral whorls.

MURA'LIS (belonging to a wall). Growing on walls.

MURA'RIUS. Synonyme for *MURALIS*.

MURICA'TED, *MURICA'TUS* (full of sharp points). Rough, with short hard tubercular excrescences.

MU'RIFORM, *MURIFOR'MIS* (*MURUS* a wall, *FORMA* shape). Flattened cellular tissue in laminæ (as in the medullary rays of exogens), and so arranged as to resemble the courses of bricks or stones in a wall.

MU'RINUS (of a mouse). Mouse-coloured; slightly reddish-grey; red with much grey.

MUSA'CEÆ. MU'SÆ (from the genus Musa). The Banana tribe. A natural order of Monocotyledones.

MUSCA'RIFORMIS (*MUSCARIUM* a fly-flap, *FORMA* shape). Brush-shaped.

MUSCA'RIUM (The top of any herb where the seeds lie). A loose and irregular corymb.

MUS'CI (*MUSCUS* Moss). The Moss tribe. A natural order of Acotyledones.

MUSHROOM-HEADED. A cylindrical body capped by a convex head of larger diameter.

MU'TICUS (beardless). Pointless.

MUTISIA'CEÆ (from the genus Mutisia). A natural group of Dicotyledones; considered as a sub-order of Compositæ; or otherwise as a distinct order.

MYCE'LIUM (μυκης a fungus). The filamentous or rudimentary body from which a fungus is developed.

MYCETOI'DEUS (μυκης a fungus, εἶδος a resemblance). With the general appearance of some fungus.

MY'CINA (μυκης a fungus). A globular stipitate Apothecium.

MY'CROPYLE. See Micropyle.

MYOPORI'NEÆ (from the genus Myoporum). A natural order of Dicotyledones.

MYRICA'CEÆ, MYRI'CEÆ (from the genus Myrica). The Gale tribe. A natural order of Dicotyledones.

MYRISTICA'CEÆ, MYRISTI'CEÆ (from the genus Myristica). A natural order of Dicotyledones.

MYRSINA'CEÆ, MYRSI'NEÆ (from the genus Myrsine). A natural order of Dicotyledones.

MYRTA'CEÆ, MYR'TEÆ, MYR'TI, MYRTI'NEÆ, MYRTOI'DEÆ (from the genus Myrtus). The Myrtle tribe. A natural order of Dicotyledones.

MYRTOI'DEUS (μυρτος the myrtle, εἶδος resemblance). Resembling the general appearance of a myrtle.

MYU'RUS (μῦς a mouse, ουρὰ a tail). Long and tapering, like a mouse's tail.

NAIADA'CEÆ, NAIA'DEÆ, NAI'ADES (from the the genus Naias). Synonymes for Fluviales.

NAIL. About half an inch in length.

NA'KED. Where a surface is destitute of pubescence. Also, where any part is exposed, or uncovered by other surrounding parts or organs.

NANDHIRO'BEÆ (from the genus Nandiroba). Synonyme for Cucurbitaceæ.

NA'NUS (*NANUS* a dwarf). Synonyme for *PUMILUS*.

NAPA'CEUS, NAPIFOR'MIS (*NAPUS* a turnip, *FORMA* shape). Turnip-shaped.

NAPOLEO'NEÆ (from the genus Napoleona: a synonyme of Belvisia). Synonymous with Belvisiaceæ, Belvisieæ. A natural order of Dicotyledones.

NARCIS'SEÆ, NARCIS'SI (from the genus Narcissus). A synonyme for Amaryllidaceæ.

NA'TURAL, *NATURA'LIS*. Used in opposition to artificial. Applied also synonymously with "Indigenous."

NAU'CUM vel *NAU'CUS* (kernel of an olive, or shell of a nut). The former has been applied to distinguish seeds whose hilum is very large. It is also used for the external fleshy covering of the stone of such fruit as the peach. The latter has been applied to certain seed vessels in the Cruciferæ where the pericarp is valveless.

NAVE-SHAPED. Like the nave of a wheel, round and depressed with a small opening, fig. 105.

105

NAVICULA'RIS (*NAVICULA* a boat). Boat-shaped.

NEBULO'SUS (Misty). Clouded.

NECESSA'RIA (*NECESSARIUS* necessary). An order of the artificial Linnean class Syngenesia, having the florets of the disk male, and those of the ray female. In other words, the capitula are monœceous.

NECK. An imaginary plane, separating the stem from the root. Also the point at which the limb separates from the sheathing petiole of certain leaves. Also the contracted portion of the calyx tube which is immediately above certain seeds surmounted by a stipitate pappus.

NECKLACE-SHAPED. Synonyme for Moniliform.

NECROG'ENUS (νεκρὸς a dead body, γεννάω to beget.) Applied to certain cryptogamic parasites, which hasten the death or destruction of the vegetables on which they live.

NECTAR, *NECTAR*. A sweetish exudation, secreted by glands in different parts of plants.

NECTARIF'EROUS, *NECTARIF'ERUS* (*NECTAR* nectar, *FERO* to bear). Secreting nectar. Also possessing a nectary.

NECTARIL'YMA (νέκταρ nectar, and ειλύω to wrap round). Any appendages which invest the nectary.

NECTARI'NUS. Something appended to a nectary.

NECTAROSTIG'MA, (νέκταρ nectar, στιγμα a spot). Some mark or depression indicating the presence of a nectariferous gland.

NECTAROTHE'CA (νέκταρ nectar, θηκη a box). Some portion of a flower immediately investing a nectariferous pore.

NECTARY, *NECTARIUM*, (*NECTAR*, nectar). Certain portions of many flowers, whether glandular or not, which cannot readily be referred to the parts forming the floral whorls Also certain parts of the whorls themselves, when they assume an anomalous character, are styled nectaries, whether they secrete nectar or not.

NEEDLE-SHAPED. Synonyme for "Acerose."

NELUMBIA'CEÆ, NELUMBO'NEÆ (from the genus Nelumbium.) A natural order of Dicotyledones; otherwise considered as a section of Nymphæaceæ.

NEMATOMY'CI (νῆμα a thread, μύκης a mushroom). Synonyme for Hyphomycetes.

NE'MEÆ (νῆμα a thread). Cryptogamic plants whose sporules elongate into a thread-like form in germination.

NE'MOBLASTUS (νῆμα a thread, βλαςός a sprout). A sporule which vegetates in the manner described under Nemeæ.

NEPENTHA'CEÆ (from the genus Nepenthes). The Pitcher plant tribe. A natural order of Dicotyledones.

NEPHROI'DEUS (Νεφρος the kidney, ειδος resemblance). Synonyme for "reniformis."

NERVA'LIS (*NER'VUS* a nerve). Synonyme for loculicidal, because the dehiscence is along the midnerve of the carpels. Also proceeding for the midrib of a leaf.

NERVA'TION, *NERVA'TIO* (*NERVUS* a nerve). The character or disposition of the nerves of a leaf or other foliaceous appendages.

NERVATO-VENO'SUS. (*NERVUS* a nerve, *VENA* a vein). When the nerves of a leaf are very much ramified and subdivided.

NERVA'TUS, NERVO'SUS. (Sinewy.) Having nerves: also, when nerves are very prominently developed

NERVE, *NERVUS.* One of the fibrous bundles of the combined vascular and cellular tissue, which extends through the parenchyma of many foliaceous organs, and often ramifies like veins or nerves in the animal structure.

NERVULO'SUS (diminutive of *NERVO'SUS*). Possessing prominent nerves.

NETTED. Synonyme for "Reticulate."

NEU'RA (νεῦρον). A "Nerve."

NEURAMPHIPET'ALÆ (νεῦρον a nerve, ἀμφι about, πεταλον a petal). Synonyme for Compositæ.

NEURO'SUS (νεῦρον a nerve). Synonyme for Nervosus.

NEUTRIFLO'RUS (*NEUTER* neuter, *FLOS* a flower.) The "ray" in the capitulum of Compositæ. When its florets are all neuter.

NI'DULANS (*NIDULATUS* nestling). Imbedded in pulp; partially encased in some covering.

NI'GER. Black. Very dark grey, but not a pure black.

NI'GRICANS. Blackish. Approaching to Niger.

NIT'IDUS (Bright). Synonyme for "lucens."

NITRARIA'CEÆ (from the genus Nitraria). A natural order of Dicotyledones.

NIVA'LIS, *NIVO'SUS* (Snowy). Flourishing in the winter season; or living amongst snow. Sometimes used as a synonyme for Niveus.

NI'VEUS (snowy). Snow-white.

NOCTUR'NAL, *NOCTUR'NUS* (of the night). Lasting through a night.

NOD'DING. When the summit is so much curved that the apex is directed perpendicularly downwards.

NODE, *NO'DUS* (a knot). Each portion of the stem from whence a leaf springs; but more especially when this portion is a little thickened or swollen.

NODO'SUS (knotty). Knotted. Also synonymous with *MONILIFORMIS*.

NODULO'SE, *NODULO'SUS* (*NODULUS* a little knot). Furnished with little knots.

NOLANA'CEÆ (from the genus Nolana). A natural order of Dicotyledones.

NOPALEÆ. Synonyme for Cactaceæ.

NOR'MAL, *NORMA'LIS* (according to rule). With reference to some principal or typical form or structure, to which other forms or structures approximate, or from which they deviate according to certain laws.

NOTA'TUS. Marked by spots, lines, &c.

NOTORHI'ZUS (*νῶτος* the back, *ῥιζα* the root). Synonyme for Incumbens, when applied to the embryo of Cruciferæ.

NUCAMENTA'CEOUS *NUCAMENTA'CEUS* (*NUCAMENTUM* the catkin of a nut). Resembling a small nut. Synonyme for Indehiscent, when applied to certain seed-vessels, as the siliquæ of some cruciferæ.

NUCAMEN'TUM (a catkin). Synonyme for *AMENTUM*.

NUCEL'LA (diminutive from Nux a nut). Synonyme for *NUCLEUS*.

NUCIFOR'MIS (*NUX* a nut, *FORMA* shape). Nearly spheroidal, but tapering at one end; i. e. shaped like a filbert.

NUCLEA'RIUS (*NUCLEUS* a kernel). The parts of a seed developed within the nucleus, viz. the embryo and albumen together.

NU'CLEOUS, *NU'CLEUS* (a kernel). The inner, pulpy, and closed sack of the ovule, within which the embryo and its immediate covering are developed, see fig. 86, to article

FORAMEN. Also, in Lichens, the disk of the shield containing the asci. Also, in Fungi, the central part of a perithecium. Also, a "clove."

NU'CULE, *NU'CULA* (a small nut). Synonyme for Glans. Also one or two forms of Apothecia peculiar to Characeæ.

NUCULA'NIUM (from *NUCULA* a small nut: because it contains hard seeds). A two or more celled indehiscent fruit, formed from a superior ovule filled with fleshy pulp, containing few or several seeds; as in the grape.

NUCULO'SUS (*NU'CULA* a small nut). Containing hard nut-like seeds.

NUDICAU'LIS (*NUDUS* naked, *CAULIS* a stalk). When a stem has no leaves.

NU'DUS. Naked.

NULLINER'VIS (*NULLUS* none, *NERVUS* a nerve). Synonyme for *ENERVIS*.

NU'MEROUS, *NUMERO'SUS*. Synonyme for Indefinite.

NUT. See Achenium.

NU'TANS. Nodding.

NUTRITION, *NUTRI'TIO* (*NUTRIOR* to nourish). The vital function by which the development of the various parts of the vegetable structure is affected.

NUX. A nut.

NYCTAGI'NEÆ, NYCT'AGINES (from the genus Nyctago of Jussieu). The Marvel-of-Peru tribe. A natural order of Dicotyledones.

NYMPHÆA'CEÆ (from the genus Nymphæa): The Water-Lily ribe. A natural order of Dicotyledones.

NYSSA'CEÆ (from the genus Nyssa). Synonyme for Santalaceæ.

OB (over against). Used in composition to signify that the point of attachment is at the opposite extremity to where it occurs in the form defined by the simple word. Thus "ob-clavate" is the inverse of "clavate," the attachment being at the thicker end. So "ob-cordate" is the inverse of "cordate," (see fig. 57) the attachment being at the narrow end.

OB-COMPRES'SUS (*OB* inversely, *COMPRESSUS* compressed). Where the compression or flattening is contrary to the more usual condition.

OB-CUR'RENS (*OB* over against, *CURRENS* running). Where the partial dissepiments in an ovarium extend to the axis, so that the capsule becomes multilocular.

Ob-imbrica'tus (*Ob* inversely, *imbricatus* imbricate). Where the imbrication is from above downwards. Also, used where rows of scales are so arranged that those on one row overtop those of the row immediately above or within them.

Obli'gulate, *Obligula'tus* (*Ob* inversely, *ligula* a strap). When the corolla of a ligulate floret (in Compositæ) is extended on the inner instead of on the outer side of a capitulum.

Obli'que (*Obliquus*). When the midrib of a plane leaf being nearly horizontal, but pointing somewhat towards the ground, the limb itself is more or less inclined to the horizon, owing to a twist in the petiole, or, in the base of the limb; see Guide, p. 64, under "Adverse." Also, when a plane leaf is so divided by the midrib that the divisions on each side are slightly unequal. See Guide, p. 79, fig. 188.

Ob'long, *Oblong'us*. Of an elliptical shape, where the major and minor axes bear a proportion to each other of about four to one. See Guide, p. 76, fig. 170.

Obo'val, *Obova'lis*. Used as a synonyme for Obovate.

Obova'te, *Obova'tus* (*Ob* inversely, *ovatus* shaped like an egg). When the point of attachment is at the narrow end of the ovate form.

Obrin'gens (*Ob* inversely, *ringens* grinning). When the ringent corolla of a floret (in Compositæ) has the interior lip composed of one-fifth, and the posterior of four-fifths, of the whole.

Obscu're (*Obscurus*). Of a dark or dingy colour.

Obstruc'tus (stopped up). Where hair, ciliæ, or other appendages, partially close the throat of a tubular corolla.

Obsub'ulate, *Obsubula'tus* (*Ob* inversely, *subula* a cobler's awl). Very narrow, pointed at the base, but gradually widening a little towards the apex.

Obsutura'lis (*Ob* over against, *sutura* a suture). Applied to the suture of a pericarp.

Obtura'tor (*Obturatus* stopped up). A small body which accompanies the pollenic masses of Orchidaceæ and Asclepiadaceæ, closing the opening of the anther.

Obtu'se (*Obtusus*). Blunt.

Obtusius'culus (diminutive of *Obtusus* blunt). Somewhat "Blunt."

Obvalla'tus (guarded strongly). When consecutive pairs

of opposite leaves are arranged at small angles of divergence from each other, and not in a brachiate manner, where the angle is a right angle.

OBVERSE, *OBVER'SUS* (turning towards). When the point of the radicle, in the seed, approaches the hilum.

OB'VOLUTE, *OBVOLUTI'VUS* (*OBVOLUTUS* muffled up, or covered over). When the margins of leaves or petals, in the bud state, are mutually enrolled one within the other. More especially applied to two plicate leaves, which, in vernation, have each one margin embraced by the folding of the other leaf, fig. 106.

106

OCEAN'IDUS (*OCEANUS* belonging to the ocean). Synonyme for Hydrophyton.

OCCULTA'TUS. Hidden.

OCELLA'TED, *OCELLA'TUS* (*OCELLUS* a little eye). Spotted in a manner somewhat resembling the pupil and iris of an eye. One spot of colour within another spot.

OCHNA'CEÆ (from the genus Ochna). A natural order of Dicotyledones.

OCHRA'CEUS (*OCHRA* ochre). The colour of yellow-ochre; yellow with a little grey.

OCHRANTHA'CEÆ (from the genus Ocranthe). A doubtful sub-order of Hypericaceæ.

OCH'REA. See *OCREA*.

OCHROLEU'CUS (ωχρολευκος of a pale yellow). With a faint tint of dingy yellow.

O'CREA (a boot). A membranous sheath at the base of some leaves, which clasps the stem. Also, (in Cyperaceæ) a membranous sheath round the base of some peduncles.

OCTAGY'NIA (ὀκτω eight, γυνη a woman). An order in the artificial system of Linneus, characterised by hermaphrodite flowers with eight pistils, or eight free styles.

OCTAN'DRIA (οκτω eight, ανηρ a man). The eighth class in the artificial system of Linneus, characterized by hermaphrodite flowers with eight stamens.

OCTAN'DROUS, *OCTAN'DER*, *OCTAN'DRICUS*, *OCTAN'DRUS*. A flower which contains eight stamens.

OCTANTHE'RUS (οκτω eight, ανθηρος an anther). Having eight fertile stamens.

OCTOGY'NIA. Synonyme for Octagynia.

OCTOGY'NICUS, *OCTO'GYNUS*. With the structure of Octogynia.

OCTO'NUS (*OCTONI* eight each). Eight together.

OCTOSTE'MONUS (οκτω eight, στημων a stamen). With eight free stamens.

OCULA'TUS (with eyes). Marked with concentric spots of different colours or tints.

OC'ULUS (the eye). The first appearance of a bud; especially the buds on a tuber. Also, a small depression on the summit of some fruits, as the Pear.

ODORA'TUS (fragrant). Possessing any decided odour, though more generally restricted to such as are sweet.

OFFICINA'LIS (*OFFICINA* a work-shop). Applied to plants which are useful in medicine or the arts.

OFF'SET. A short postrate branch whose terminal bud takes root, but which does not (as in the Runner) branch again in a similar manner.

OF'TEN-BEARING Producing more than twice in one season.

OLACA'CEÆ, OLACI'NEÆ (from the genus Olax). A natural order of Dicotyledones.

OLEA'GINOUS, *OLEA'GINUS* (of an olive tree). Succulent and oily. Also, like oil.

OLEA'CEÆ, OLEI'NEÆ (from the genus Olea). The Olive tribe. A natural order of Dicotyledones.

O'LENS (smelling strong). Strong-scented, whether agreeable or nauseous.

OLIVA'CEUS (*OLIVA* an olive). Of an olive-green colour. Orange and grey.

OLIVÆ'FORMIS, *OLIVIFOR'MIS* (*OLIVA* an olive, *FORMA* shape). Shaped like an olive. Ellipsoidal.

OMOPLE'PHYTUM (ομοπλεκὴς folded together, φυτον a plant). Synonyme for a plant with a monadelphous flower; because the stamens are blended together into one bundle.

OMPHALO'DIUM (ομφαλος the navel, εἶδος a resemblance). The mark left in the hilum by the passage of the vessels of the raphe.

ONAGRA'CEÆ, ONAG'RÆ, ONAGRA'RIÆ (from Onagra, a synonyme for the genus Œnothera). The Evening Primrose tribe. A natural order of Dicotyledones.

ONE-RIBBED. When there is only one prominent nerve or midrib in a leaf.

ONE-SIDED. Synonyme for "Secund."

OPAQUE (*OPACUS*). When the *surface* is dull, or not at all shining.

OPER'CULATE (*OPERCULATUS* covered by a lid). Closed by an "operculum."

OPER'CULUM (a cover or lid). The expansion at the extremity of a "pitcher," which closes its mouth. The superior portion of the theca of Mosses, which, on their ripening, generally becomes detached and exposes the peristome.

OPER'TUS (shut up). Synonyme for "*TECTUS*."

OPHIOGLOSSA'CEÆ, OPHIOGLOS'SEÆ (from the genus Ophioglossum). A natural order of Acotyledones, of the group Filices, or Ferns.

OPHIOSPER'MEÆ (οφις a snake). Synonyme for Myrsinaceæ; because the embryoes of some of the species are like a small snake in appearance.

OPLA'RIUM (οπλη a hoof). Synonyme for "Scyphus."

OP'POSITE (*OPPOSITUS* set against). When similar parts or organs are so arranged in pairs, that one of them is immediately on the opposite side of some interposed body, or of the axis about which they are disposed. Thus two leaves may be opposite upon the stem; fig. 107, two petals about the pistil, &c. Also, a synonyme for "Adverse." Also, (*OPPOSITIVUS*) when some part or organ stands immediately before another, in contradistinction to an "alternate" arrangement.

OPPOSITIFLO'RUS (*OPPOSITUS* opposite, *FLOS* a flower). Where the penduncles are "opposite."

OPPOSITIFO'LIUS (*OPPOSITUS* opposite, *FOLIUM* a leaf). Where the leaves are opposite. Also, when any other organ, as a tendril, or the inflorescence, is oppositely arranged with respect to a leaf.

ORANGE. Yellow and red in about equal proportions.

ORBIC'ULAR, *ORBICULA'RIS* (*ORBICULATUS* of a round form). Perfectly or very nearly circular.

ORBIC'ULUS (a little round ball). A description of fleshy "*corona*" surrounding the organs of fructification in the genus Stapelia. Also a round flat hymenium in certain fungi.

ORBIL'LA (diminutive of *ORBIS* an orb). Synonyme for certain "scutella" of lichens.

ORCHIDA'CEÆ, ORCHI'DEÆ, OR'CHIDES (from the genus Orchis). The Orchis tribe. A natural order of Monocotyledones.

ORCHIDA'CEUS. Furnished with two tubercles at the root, like those in many species of Orchis.

Organ (*Organum*, some machine or engine), A general name for any defined subordinate part of the vegetable structure, external or internal; as cell, fibre, leaf, root, &c.

Orgy'a (*οργυια* from five to six feet). A toise.

Orgya'lis. The length of a toise.

Orobancha'ceæ, Oroban'cheæ, Oroban'chinæ (from the genus Orobanche). The Broom-rape tribe. A natural order of Dicotyledones, composed of certain plants which grow parasitically on the roots of others.

Orthoplo'ceæ (ὁρθὸς straight, πλεκω to fold together). A section of Cruciferæ, characterized by a modification in the "Incumbent" arrangement of the parts of the embryo. The cotyledons are folded longitudinally so as partially to embrace the radicle; fig. 108. (*a*) is the seed, (*b*) the same cut transversely. It is synonymous with "Conduplicatæ."

Ortho'tropal, Ortho'tropous, *Ortho'tropus* (ορθὸς straight, τρεπω to turn). Where the embryo (*e*) is straight, fig. 109; but so lies in the seed that the radicle is towards the hilum (*h*), owing to the inversion of the nucleus (*n*). This term is also applied to the entire ovule or seed, without reference to this position of the embryo; when the nucleus is straight, and the chalaze (*c*) and hilum (*h*) correspond or are close together; and consequently where the direction of the embryo is "antitropal," or the very reverse of that here described.

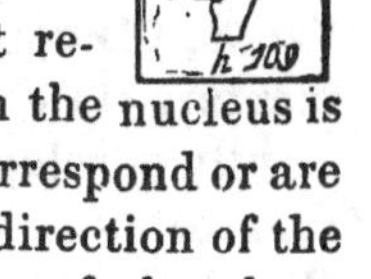

Oscilla'nus, Oscillato'rius (*Oscillum* an image hung up to swing about). Synonyme for Versatilis.

Osmunda'ceæ (from the genus Osmunda). A natural order of Acotyledones, belonging to the section of Ferns.

Os'seous (*Osseus* made of bone). Synonyme for "Bony."

Ossic'ulus (diminutive of *os* the "stone" in fruits). A "stone" in fruit. Synonyme for Pyrena.

Ostari'phytum (ὀσταριον a little bone, φυτον a plant). A plant which bears a "Drupe"

Osti'olum (a little door). The orifice of the perithecium and apothecium, in the lower tribes of Acotyledones.

Ostiola'tus. Furnished with *Ostiola*.

Osyri'deæ, Osyri'næ (from the genus Osyris). Synonyme for Santalaceæ.

Out'line. The figure obtained by circumscribing a surface 59.

in a continuous line, without reference to marginal indentations or inequalities.

O'VAL, *OVA'LIS* (*OVATUS*). Elliptical. Where the major and minor axes bear the ratio of about two to one to each other. Fig. 110.

O'VARY, *OVA'RIUM* (*OVUM* an egg). The lowermost portion of the pistil, containing the ovules; and which ultimately becomes the fruit.

OVA'TE, *OVA'TUS* (shaped like an egg). Of the form of an egg, when applied to a solid body; but when applied to a superficial area, it means the figure presented by a longitudinal section of an egg, broader at the base than at the apex. Sometimes used synonymously with Oval. Fig. 111.

OVEL'LUM. Synonyme for *OVULUM*.

OVIFOR'MIS (*OVUM* an egg, *FORMA* shape). Synonyme for Ovate.

OVO'ID, OVOI'DAL, *OVOI'DEUS*, *OVULA'RIS* (*OVUM* an egg, ειδος a likeness). Synonyme for Ovate, when applied to a solid form.

OVULA'TUS. Approaching the Ovoid form.

O'VULE, *O'VULUM* (diminutive for Ovum an egg). The rudimentary state of a seed, so long as the ovary is not sufficiently developed to be styled the fruit.

OXALIDA'CEÆ, OXALI'DEÆ (from the genus Oxalis). The Wood Sorrel tribe. A natural order of Dicotyledones; otherwise considered to be a tribe of Geraniaceæ.

OXYACAN'THUS (ὀξυς sharp, ἄκανθα a spine), Furnished with many sharp thorns or prickles.

OXYCAR'PUS (οξυς sharp, καρπος fruit). Where the fruit is sharp-pointed.

PACHYCAR'PUS (παχυς thick, καρπος fruit). Where the pericarp is very thick.

PAGI'NA (a page of a book). The upper or under surface of flat leaves.

PAINTED. Marked in streaks, by various tints, of different colours.

PAIRED. When the leaflets of a pinnate leaf are only two in number.

PALA'CEOUS, *PALA'CEUS* (*PALA* the enchasing or setting round a stone in jewelry). When the edges of any organ, but more especially the margins of a leaf, adhere to the support.

PAL'ATE, *PALA'TUM* (the roof of the mouth). The inferior surface of the throat in ringent and personate corollas, where it is elevated into two longitudinal ridges, with a depression between them.

PA'LEA (chaff). One of the membranaceous bracts which forms the perianth of the Grass tribe. The chaff-like scaly bracts on the receptacle of certain Compositæ, seated at the base of the florets; also the bracts of the involucrum in the same Order.

PALEA'CEOUS (*PALEA* chaff). Chaffy,

PALE, *PAL'LIDUS*. With very slight tinge of colour.

PALE'OLA (diminutive of *PALEA*). Synonyme for Glumella.

PALM, *PALMA'RIS* (breadth of four fingers), *PALMUS* (from thumb to little finger). About three inches in length.

PALMA'CEÆ, PALMÆ (*PALMA* the Date). The Palm tribe. A natural order of Monocotyledones.

PAL'MATE, *PALMA'TUS* (marked by the palm of the hand). When the arrangement of subordinate parts of any organ is such as to imitate the form of an open hand; fig. 112.

PALMAT'IFID, *PALMATIF'IDUS* (*PALMATUS* and *FINDO* to cleave). When the subdivisions of a simple leaf, having a palmate arrangement, extend about half way towards the base.

PALMA'TILOBED, *PALMATILOBA'TUS* (*PALMATUS* and *LOBATUS*). Where the lobes of a simple leaf have a palmate arrangement.

PALMA'TIPAR'TITE, *PALMA'TIPARTI'TUS* (*PALMATUS* and *PARTITUS*). When the subdivisions of a simple leaf, having a palmate arrangement, extend considerably more than half way to the base.

PALMA'TISEC'TUS (*PALMATUS*, and *SECO* to cut). When the subdivisions extend deeper than the *PALMATIPARTITE* arrangement, almost to the very base.

PANDANA'CEÆ, PANDA'NEÆ (from the genus Pandanus). The Screw-pine tribe. A natural order of Monocotyledones.

PANDU'RIFORM, *PANDU'RÆFORMIS*, *PANDU'RIFORMIS*, *PANDURA'TUS* (*PANDURA* a musical instrument, *FORMA* shape). Where an oblong or obovate leaf is indented in the lower half by a deep sinus; fig. 113.

PAN'ICLE, *PANIC'ULA* (the down upon reeds). An inflorescence (fig. 114) where the rachis either subdivides into several branches, or is furnished with distinct branching peduncles.

PAN'ICLED, *PANICULA'TUS*. Where the flowers are arranged in a panicle.

PANNEXTER'NA. Synonyme for Epicarpium.

PANNIFOR'MIS, *PANNO'SUS* (*PANNUS* cloth, *FORMA* shape). Looking like a piece of cloth; somewhat thick and spongy, as the thallus of certain Lichens.

PANNINTER'NA. Synonyme for Endocarpium.

PAPAVERA'CEÆ (from the genus Papaver). The Poppy tribe. A natural order of Dicotyledones.

PAPAYA'CEÆ, PAPA'YÆ (from the specific name of the Carica Papaya). The Papaw tribe. A natural order of Dicotyledones.

PA'PERY. Having the consistency of letter paper. A common character with leaves.

PAPILIONA'CEOUS, *PAPILIONA'CEUS* (*PAPILIO* a butterfly). An irregular corolla composed of five petals (fig. 115), one of which is called the "standard" (*a*); two others, placed laterally, are the "wings" (*b*); and two (opposite the standard, and more or less cohering) form the "keel" (*c*).

PAPIL'LA (the nipple). A small elongated protuberance, formed of a distended cell of the cellular tissue, upon various surfaces.

PAPILLA'RIS. Resembling a papilla, but of larger dimensions, and composed of several cells.

PAPILLA'TUS, *PAPILLIF'ERUS*, *PAPILLO'SUS* (*PAPILLA* the nipple, *FERO* to bear). Covered with papillæ.

PAPPIF'ERUS (*PAPPUS* thistle-down, *FERO* to bear). *PAPPO'FERUS*, *PAPPO'SUS* (παπποϛ thistle-down, φερω to bear). Furnished with pappus.

PAP'PUS (thistle-down). The peculiar limb to the calyx of the florets of Compositæ; which is very frequently of a downy texture, as in Thistles, &c.

PA'PULA (a pimple). An elongated papilla.

PAPULIF'ERUS, *PAPULO'SUS*. Covered with papulæ.

PAPYRA'CEUS, *PAPYRIF'ERUS*, (made of *PAPYRUS*). Papery.

PARABOL'IC, PARABOL'ICAL, *PARABOL'ICUS* (from the mathematical figure termed a Parabola). A synonyme for Ovato-oblong.

Paracar'pium (παρα about, καρπος fruit). An abortive ovary. Also a persistent portion of some styles or stigmas.

Paracorol'la (παρα about, *corolla* the corolla). Any appendage to a corolla that is usually classed among nectaries.

Parallel, *Paralleli'cus*, *Parallel'us*. Where the axes of two parts lie parallel to each other.

Paralleliner'vis, *Parallel'inervius* (*Parallelus* parallel, *nervus* a nerve.) Straight ribbed.

Parallelivenо'sus (*vena* a vein). Synonyme for Parallelinervis.

Parapetaloid', *Parapetaloi'deus.* Bearing a Parapetalum.

Parapet'alum (παρα about, πεταλον a petal). Synonyme for Peripetalum.

Paraphyl'lum (παρα about or nearly, φυλλον a leaf). Certain foliaceous expansions on some calyces. A synonyme for Stipule.

Para'physis (παρα about, φυω to grow). Succulent and jointed filaments intermixed with (and considered to be abortive) thecæ in Mosses. Also the rays which form the corona in the genus Passiflora have been called Paraphyses and Parastades.

Paraste'mon (παρα about, στημων a stamen). A nectariform appendage to the stamens.

Par'asite, *Parasi'ta*. A plant which obtains its nourishment directly from the juices of some other plant to which it is attached.

Paren'chyma (παρεγχυω to strain through). Masses of cellular tissue; more especially restricted to those in which the cells are polygonal, and not fusiform.

Pa'ries (a wall). The inner surface of the pericarp; or of a tubular calyx.

Pari'etal, *Parieta'lis* (*Paries* a wall). Attached to the "paries."

Pari-pinna'tus (*Par* equal, *penna* a wing). Synonyme for "abruptly pinnate."

Paronychi'eæ (from the genus Paronychia). Synonyme for Illecebraceæ.

Par'ted, Par'tite, *Parti'tus* (proportionably divided). Subdivided into similar segments, the divisions extending nearly to the base; fig. 116.

Par'tial, *Partia'lis* (*Pars* a part). A subordinate part in some general arrangement; thus, each leaflet of a compound leaf has its "partial" petiole attached to a main petiole. See, also, "general," as applied to the Umbel.

PARTI'BILIS (*PARS* a part). Where one portion, arrived at maturity, separates spontaneously from another portion.

PARTI'TION, *PART'ITIO* (a dividing). The segments of a partite leaf,

PARTITIONED. Subdivided into cells.

PAR'VUS (small). Applied relatively, where some object is small by comparison with similar objects. Thus, *PARVIFLORUS*, *PARVIFOLIUS*, &c., are terms given to plants whose flowers or leaves are smaller than those of other allied species.

PAS'SALUS (πασσαλος a button). An unusual term for a gamosepalous calyx.

PASSIFLORA'CEÆ, PASSIFLO'REÆ (from the genus Passiflora). The Passion-flower tribe. A natural order of Dicotyledones.

PATEL'LA, *PATEL'LULA* (a deep dish with a broad rim). An orbicular sessile apothecium with a marginal rim distinct from the thallus.

PATELLIFOR'MIS (*PATELLA* a dish, and *FORMA* shape). Knee-pan-shaped.

PA'TENT, *PA'TENS* (opened, extended). Spreading.

PA'TULUS (spreading). Slightly spreading.

PAU'CUS (few). Applied relatively, when certain portions are few in one species compared with similar portions in an allied species. Thus, *PAUCI-FLORUS*, *PAUCI-FOLIUS*, and *PAUCI-JUGATUS*, &c., are used with respect to flowers, leaves, leaflets, &c.

PEAR-SHAPED. Ovate beneath and conical above.

PEC'TEN (a comb). Synonyme for Trichidium.

PECTINA'TE, *PECTINA'TUS* (*PECTEN* a comb). Where a "pinnatifid" incision has the segments parallel, narrow, and close, like the teeth in a comb. Fig. 117.

117

PEDALIA'CEÆ, PEDALI'NÆ (from the genus Pedalium). The Oil seed tribe. A natural order of Dicotyledones.

PEDA'LINERVED, *PEDA'LINERVIS* (*PEDALIS* a foot long, *NERVUS* a nerve). A form of nervation in leaves where the mid-nerve stops short, but where two strong lateral nerves are produced at its base, and from these originate others which extend only towards the apex.

PEDA'LIS (a foot long). About a foot in length.

PE'DATE, *PEDA'TUS* (having feet). Where the subordinate parts have a palmate arrangement, with the addition of further subdivisions in the lateral portions. See fig. 118.

PEDAT'IFID, *PEDATI'FIDUS* (*PEDATUS* and *FINDO* to cleave). Where the sub-divisions of a simple leaf, arranged pedately, extend about half way towards the base.

PEDA'TILOBED, *PEDA'TILOBUS* (*PEDATUS* and *LOBATUS*). Where the lobes of a leaf are arranged Pedately.

PEDA'TINERVIS. Synonyme for *PEDALINERVIS.*

PEDA'TI-PARTITE, *PEDATIPARTI'TUS* (*PEDATUS* and *PARTITUS*). Where the subdivisious of a simple leaf are arranged pedately, and extend considerably more than half way towards the base.

PEDA'TISECTUS (*PEDATUS* and *SECO* to cut). When the subdivisions (as in Pedati-partite) extend very nearly to the base,

PED'ICEL, *PEDICEL'LUS* (*PES* a foot). One of the partial flower stalks, or immediate supports of a flower in an inflorescence composed of flowers arranged upon a main peduncle.

PEDICEL'LATE, *PEDICELLA'TUS* (*PES* a foot). Furnished with a pedicel.

PEDICEL'LULUS (diminutive of *PEDICELLUS*). A filiform support to the ovary in certain Compositæ.

PEDICULA'RES (from the genus Pedicularis). A Synonyme for Scrophulariaceæ.

PED'ICULUS (stalk of Apples and other fruit). A thin and somewhat long support to any organ. A synonyme for the filament.

PEDI'FERUS (*PES* a foot, *FERO* to bear). Furnished with a small support or stalk.

PEDILA'TUS. Furnished with a Pedilis.

PEDI'LIS (*PES* a foot). The contracted upper portions of the calyx-tube in such florets of Compositæ as furnish a stipitate pappus; fig. 119.

PE'DUNCLE, *PEDUN'CULUS* (*PES* a foot). The main stalk or support to the inflorescence; more especially when this is limited to a solitary flower.

PEDUNCULA'TE, *PEDUNCULA'RIS*, *PEDUNCULA'TUS*, *PEDUNCULO'SUS.* Furnished with a peduncle. Also, *PEDUNCULA'RIS* and *PEDUNCULEA'NUS* from a modified state of the peduncle.

PEL'LICLE, *PELLIC'ULA* (a little skin). An extremely delicate superficial membrane. A synonyme for epidermis.

PELLICULA'RIS. With the character of a pellicle.

PELLU'CID, *PELLU'CIDUS* (transparent). Perfectly or only partially transparent.

PELO'RIA (πέλωρ a monster). A form assumed by certain flowers, which being unsymmetrical in their usual state, become symmetrical in what may be considered as a return to their normal type.

PEL'TA (a target). A flat apothecium without a rim.

PELTA'TE *PELTA'TUS* (armed with a pelta). Where a support is inserted at some distance within the margin, and is not in the same plane as the flat surface which rests upon it. Fig. 120.

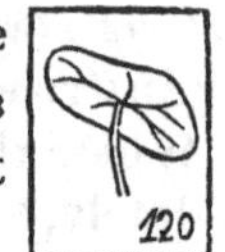
120

PELTI'DEUS, PELTIFOR'MIS (*PELTA* a target, *FORMA* shape). When the pileus of an agaric, or the apothecium of a Lichen, &c., is orbicular, or buckler-shaped.

PEL'TINERVED, *PELTINER'VIS* (*PELTA* a target, *NERVA* a nerve). Where the nerves diverge from the summit of a petiole, and spread on all sides in a plane that is much inclined to it.

PELTOI'DEUS (πέλτη a shield, εἶδος like). Shaped like a shield; somewhat orbicular, and convex on the upper side.

PENÆA'CEÆ (from the genus Penæa). A natural order of Dicotyledones.

PENDENT, (*PENDENS* hanging from). So much inclined that the apex is pointed vertically downwards.

PENDULOUS, (*PENDULUS* hanging down, *PENDULINUS*). Synonyme for "Pendent"; more especially where the flexure arises from the weakness of the support.

PENICILLA'TUS (*PENICILLUS* a painter's pencil). Bordered or tipped with fine hairs resembling those of a hair pencil.

PENICILLIFOR'MIS (*PENICILLUS* a painter's pencil, *FORMA* shape). When hairs, resembling those of a hair pencil, are collected together in the shape of one.

PENNATICI'SUS (*PENNATUS* winged, *CÆDO* to cut). When the incisions of leaves are disposed in a pinnate manner.

PEN'NATE, *PENNA'TUS, PENNAT'IFID*, &c. Synonymes for Pinnate, Pinnatifid, &c.

PENNIFOR'MIS (*PENNA* a feather, *FORMA* shape). With a pinnate arrangement of subordinate parts.

PENNINER'VED, *PENNINER'VIS, PENNINER'VIUS* (*PENNA* a

feather, *NERVA* a nerve). Where the primary nerves of a leaf are straight, and diverge from the midrib in a pinnate manner.

PENNIVE'NIUS (*PENNA* a feather, *VENA* a vein). Synonyme for Penninervius.

PENTACHAI'NIUM, *PENTAKE'NIUM* (πεντε five, α not, χαινω to open). A five-celled fruit, otherwise resembling a cremocarp.

PENTAGY'NIA (πεντε five, γυνη a woman). An order in the Artificial System of Linneus, characterized by flowers with five pistils, or five distinct styles.

PENTAGY'NOUS, *PENTA'GYNUS*. Possessing the structure explained under Pentagynia.

PENTAN'DRIA (πεντε five, ανηρ a man). An artificial class in the Linnean System, characterized by hermaphrodite flowers with five stamens. Also, an order in some classes characterized by other considerations than the mere number of the stamens.

PENTAN'DROUS, *PENTAN'DRUS*. Possessing the structure explained in Pentandria.

PENTARRHI'NUS (πεντε five, αρρην male). Synonyme for Pentandrous.

PE'PO, *PEPONI'DA*, *PEPO'NIUM*. A gourd.

PEPONI'FERÆ. A group of Dicotyledones, including the orders whose fruit is a gourd.

PERAPE'TALUM (περι about, πεταλον a petal). Any appendage to a petal. A synonyme for Nectarilyma and Parapetalum.

PERAPHYL'LUM (περι about, φυλλον a leaf). Synonyme for Paraphyllum.

PEREM'BRYUM (περι about, εμβρυον the embryo). The portion of monocotyledonous embryos investing the plumule and radicles not externally distinguishable.

PEREN'NANS. Synonyme for Perennis, and also for Persistens.

PERENNIAL, *PEREN'NIS* (durable).

PERFECT, *PERFEC'TUS*. Where every part of a flower is developed and none abortive.

PERFOLIATE, *PERFOLIA'TUS* (*PER* through, *FOLIUM* a leaf). When leaves, oppositely arranged, unite at their bases (*a*); or when the basal lobes of clasping leaves become united, so that the axis about which they are placed appears to pass through them (*b*); fig. 121.

PERFORA'TED, *PERFORA'TUS* (bored through). Pierced with one or more holes.

PERFOS'SUS (dug through). Synonyme for Perfoliatus.

PERFU'SUS (covered all over). Completely covered by.

PERGAMENTA'CEUS (*PERGAMENA* parchment). Having the consistency of parchment.

PERIAN'DRICUS (περι around, ανηρ a man). When a nectary is ranged around the stamens.

PERIAN'TH, *PERIAN'THIUM* (περι around, ανθος a flower). The external floral whorl or whorls which surround the stamens and pistil. In this sense it includes calyx and corolla. Linneus applied it to calyx and involucrum. It has been restricted to the upper portions of epigynous calyces; and also used synonymously with Periclinium and Periphoranthium.

PERIAN'THEUS. Possessing a perianth.

PERIANTHIA'NUS. Possessing a single investing perianth.

PERICALY'CIUS (περι around, καλυξ a calyx). Synonyme for Peristamineus

PER'ICARP, *PERICAR'PIUM* (περι around, καρπος fruit). The ripened condition of the ovary or ovaries, and any externally adhering appendages of the flower. A synonyme for the theca of mosses.

PERICAR'PICUS. Belonging to the pericarp. When (as respects their axis) seeds have the same general direction as the pericarp to which they are attached.

PERICEN'TRICUS (περι around, κεντρον the centre). Where perigynous stamens are arranged concentrically with the calyx.

PERICHÆTIAL, *PERICHÆTIALIS.* Bracts forming the Perichætium.

PERICHÆ'TIUM (περι around, χαιτη the mane or bristle) The leafy involucrum surrounding the seta of mosses.

PERICLA'DIUM (περι around, χλαδος a branch). The lowermost clasping portion of sheathing petioles.

PERICLI'NIUM (περι around, κλινη a bed). The involucrum of Compositæ.

PERICOROLLA'TUS (περι around, *COROLLA* a corolla). A Dicotyledonous flower with a monopetalous perigynous corolla.

PERI'DIOLUM (diminutive for Peridium). Used either synonymously with Peridium, or more particularly for the internal coat, where more than one is present in the Peridium.

PERI'DIUM (περιδεω to wrap round). The coat immediately enveloping the sporules of the lower tribes of Acotyledones.

PERI'DROMA (περι around, δρομος a course). Synonyme for the Rachis of Ferns.

PERIEN'CHYMA (περι around, εγχυμοσ succulent). Irregular cellular tissue, chiefly in glands and spheroidal masses.

PERIGO'NE, *PERIGO'NIUM* (περι around, γονευς a parent). A synonyme for Perianth; but more especially when reduced to a single floral whorl, possessing a character intermediate between calyx and corolla. Synonyme for Perichætium.

PERIGONIA'RIUS. With the character of a Perigone. Double flowers resulting from transformation or multiplication of the parts of the floral whorls, assuming the character of the parts of the Perianth.

PERIGYNAN'DRA, *PERIGYNAN'DRUM* (περι around, γυνη a woman, ανηρ a man). Synonyme for Perianthium.

PERIGY'NIUM (περι around, γυνη a woman). The membranous perianth of Carices; and the hypogynous set found in other Cyperaceæ.

PERIGY'NOUS, *PERIGY'NUS* (περι around, γυνη a woman). Where the ovarium is free, but there exists adhesion between the stamens or corolla with the calyx. *PERIGYNICUS* applies, also, where there is a partial adhesion between the ovary and calyx, but not so high up as with the stamens.

PERIPETA'LIA (περι around, πεταλον a petal). Polypetalous dicotyledones with perigynous stamens.

PERIPE'TALOUS, *PERIPE'TALUS* (περι around, πεταλον a petal). Where nectaries surround the corolla.

PERIPHE'RICUS (περι around, φερο to bear). When a long embryo is so curved as to extend within the whole circumference of the seed, till the two extremities are brought close together; fig. 122.

PERIPHORAN'THIUM (περι around, φέρο to bear, ανθος a flower). The involucrum of Compositæ. Synonyme for Perianthium.

PERIFO'RIUM (περι around, φερο to bear). When the immediate support to the ovary is somewhat fleshy and elongated; having the corolla and stamens attached to it.

PERIPHYL'LUM (περι around, φυλλον a leaf). Synonyme for Glumella.

PERIPO'DIUM (περι around, πους, ποδος a foot). Synonyme for Perichætium.

PERIPTERA'TUS, *PERIP'TERUS* (περι around, πτερον a wing). Surrounded by a thin membranous expansion, or wing.

PERISPERM, *PERISPER'MIUM*, *PERISPER'MUM* (περι around, σπερμα a seed). Synonyme for albumen. Used also for the innermost envelope to the seed.

PERISPERMA'TUS. Furnished with perisperm.

PERISPER'MICUS. When the perisperm is reduced to a simple lamina; or when the seed is not furnished with a true perisperm. A synonyme for Perispermatus.

PERISPORA'NGIUM (περὶ around, σπορα a seed, αγγειον a vessel). Synonyme for the indusium of ferns.

PERISPO'RIUM *PERISPO'RUM* (περι around, σπορα a seed). Synonyme for Perigynium.

PERISTA'CHYUM (περι around, σταχυς the ear of corn). Synonyme for Gluma.

PERISTAMI'NIA, *PERISTE'MONES*. A petalous dicotyledon with perigynous stamens.

PERISTOME, *PERISTO'MA*, *PERISTO'MIUM* (περι around. στομα) A simple or double membrane, closing the mouth or opening, in the theca of Mosses; and, after the lid (or operculum) has been removed in the ripe state, most usually sub-dividing into a definite number of laminæ, termed teeth; this number being generally some multiple of four.

123.

PERISTOMA'TUS, *PERIS'TOMUS*. Furnished with a Peristome.

PERISTO'MICUS (περι around, στομα the mouth). When perigynous stamina are attached round the mouth of the tube of the calyx.

PERISTY'LICUS (περι around στυλος the style). Where epigynous stamens are inserted (or arise from) between the style and limb of the calyx.

PERI'SYPHE (περι around, συσφαιροω to make round). Synonyme for Perichætium.

PERITHE'CIUM (περι around, θηκα a box). The envelope surrounding the masses of fructification in some Fungi and Lichens.

PER'ITROPAL, *PER'ITROPUS*, (περι around, τρεπω to turn). Where the axis of a seed is perpendicular to the axis of the pericarp, to which it is attached.

PERLA'RIUS, *PERLA'TUS*. (*PERLA* a pearl). Shining with a pearly lustre. Furnished with rounded tubercular appendages

PER'MANENT, *PER'MANENS*. Synonyme for Persistent.

PEROCI'DIUM. Synonyme for Perichætium.

PERONA'TUS (wearing high shoes). When the stipes of Fungi are thickly clothed with woolly matter, becoming powdery.

PERPENDIC'ULAR (*PERPENDICULARIS*). When an organ maintains, naturally, a vertical direction with respect to the horizon; or with respect to some other part to which it is attached.

PERPUSIL'LUS. Synonyme for Pusillus.

PERSIS'TENT (*PERSISTENS*, remaining). When any part, or organ, in particular plants, remains beyond the period of maturity and fall, appointed for similar parts in other plants; or to the period of full maturity attained by neighbouring parts.

PERSONA'TÆ (from the personate character of some of the flowers). A synonyme for Scrophulariaceæ.

PER'SONATE, *PERSONA'TUS* (*PERSONA* a mask). A form of monapetalous bilabiate corolla, where the orifice of the tube is closed by an inflated projection of the throat; fig. 124.

PERTU'SUS (bored through). Synonyme for Perforatus.

PER'VIOUS, *PER'VIUS* (having a way through). Admitting the passage of some portion without interruption; as where the pith is continued through a node in the stem.

PER'ULA (a little pouch). A sac formed in some Orchideæ by the prolonged and united bases of two of the segments of their perianth. The cap-like covering of buds, formed by the abortion of their outer leaves.

PES. A foot.

PETAL, *PE'TALUM* (πεταλον an unfolded leaf). One of the foliaceous expansions constituting the subordinate parts of the floral whorl, termed the corolla.

PETALA'TUS. Possessing a corolla.

PE'TALIFORMIS (*PETALUM* a petal, *FORMA* shape). Synonyme for Petaloideus.

PETALI'NUS. Belonging to a petal.

PETAL-LIKE, PETAL-SHAPED. Synonyme for Petaloid.

PETALO'DEUS. Where a double flower is formed by the transformation of any of the parts about the corolla into petal-like expansions. Also plants furnished with petals.

PETALOID, *PETALO'IDEUS* (πεταλον a petal, ειδος likeness). Assuming the more usual character of the petals of flowers

—that is to say, of thin, membranous, and coloured foliaceous expansions.

Petaloma'nia (πεταλον a petal, μανια madness). A monstrous development of parts in the flower, not belonging to the corolla, into petal-like expansions.

Petaloste'mones (πεταλον a petal, στημον a stamen). Plants bearing flowers with their stamens adhering to the corolla.

Petiola'ceus, *Petiola'ris*. Having reference to the petiole; either by attachment to it, or by its transformation, or some remarkable appearance, &c.

Petiola'ted, *Petiola'tus*. Furnished with a petiole.

Pet'iole, *Peti'olus* (a fruit-stalk, a little foot). The stalk or support by which the blade, or limb of a leaf, is attached to the stem.

Petiolea'nus. Originating in some modification of a petiole.

Petiolula'ris. Applied to the stipules of compound leaves which are attached to the petiole.

Petiveria'ceæ (from the genus Petiveria). A natural order of Dicotyledones.

Phænocar'pus (φαινω to appear, καρπος fruit). When a fruit is very distinct, the pericarp having contracted no adhesion with surrounding parts.

Phænog'amous, *Phænog'amus* (φαινω to appear, γαμος marriage). Synonyme for Phanerogamous.

Phalarsi'phytus (φαλαγξ a regiment, αρρην male, φυτον a plant). Synonyme for Polyadelphus.

Phaneran'thus (φανερος apparent, ανθος a flower). Where the flower is manifest.

Phaneran'therus (φανερος apparent, ανθηρος an anther). Where the anthers protrude beyond the perianth.

Phanerog'amous, *Phanerog'amus* (φανερος apparent, γαμος marriage). Plants in which the stamens and pistils are distinctly developed; including the two classes of Dicotyledones and Monocotyledones.

Philadelpha'ceæ, Philadel'pheæ (from the genus Philadelphus). The Syringa tribe. A natural order of Dicotyledones.

Phœni'ceous, *Phœni'ceus* (purple). Red, very slightly tinged with grey.

Phoran'thium (φερω to bear, ανθος a flower). A synonyme for Clinanthium.

Phragmi'gerus (φραγμα a fence, *gero* to bear). Where an

otherwise continuous cavity is separated into subordinate parts by transverse partitions of a membranous or cellular character.

PHYCO'MATER (φυκος a sea-weed, ματηρ a mother). The gelatinous matter investing the sporules of certain Algæ.

PHYCOSTE'MON (φυκοω to paint or disguise, στημων a stamen). Synonyme for Disk, as applied to various nectariferous expansions among the floral whorls.

PHYLLOCLA'DIUM (φυλλον a leaf, κλαδος a branch). Synonyme for Phyllodium.

PHYLLODI'NEUS. Bearing Phyllodia.

PHYLLO'DIUM (φυλλον a leaf, ειδος form). When the limb of a leaf is abortive, and the petiole becomes developed into a flattened expansion. Also a synonyme for Phylloma.

PHYLLOI'DEUS (φυλλον a leaf, ειδος resemblance). Synonyme for Foliaceous.

PHYLLOLOBEÆ (φυλλον a leaf, λοϐος a lobe). An extensive group among the Leguminosæ, characterized by the cotyledons being foliaceous.

PHYL'LOMA (φυλλον a leaf, ομας an assemblage). The aggregate mass of germs contained in a leaf bud.

PHYLLOMA'NIA (φυλλον a leaf, μανια madness). Where a superabundance of leaf-buds is formed in comparison with flower-buds.

PHYLLOTA'XIS (φυλλον a leaf, ταξις order). The manner in which leaves are arranged upon the stem.

PHYMATO'DEUS (φυμα a wart, ειδος resemblance). Synonyme for Verrucosus.

PHYSE'MA, *PHYSEU'MA*, (φυσημα a swelling). Synonyme for Frons in the aquatic Algæ.

PHYTOG'RAPHY, *PHYTOGRA'PHIA* (φυτον a plant, γραφω to write). That department of botany which includes the description of plants.

PHYTOLACCA'CEÆ, PHYTOLAC'CEÆ (from the genus Phytolacca). The Virginian Poke tribe. A natural order of Dicotyledones.

PHYTOL'OGY, *PHYTOLO'GIA* (φυτον a plant, λογος a discourse). Synonyme for Botany.

PHYTONO'MIA (φυτον a plant, νομος a law). Synonyme for Botanical Physiology.

PHYTONY'MIA (φυτον a plant, ονυμα a name). Synonyme for Organographia.

Phytotro'phia (φυτον a plant, τροφη nourishment). The science of plant-culture.

Pi'ceus (pitch-black). Black, with a brownish tinge. Red, almost concealed by admixture with intensely deep grey.

Pic'tus. Painted.

Pila'ris. Used as a synonyme for Pilosus.

Pile'ola, (*Pile'olum* a little cap). A primordial leaf, like an elongated extinguisher, which completely encloses all the gemmule.

Pi'leus (a cap). The expanded upper portion of the stipes of certain fungi, in which the sporules are embedded.

Pili'dium (πιλος a cap, ειδος resemblance). An hemispherical apothecium in some Lichenes, the surface of which changes to powder.

Pilif'erus (*Pilus* hair, *fero* to carry). Bearing hair on the surface. Also, synonyme for Hair-pointed.

Pilo'se, *Pilo'sus*. Synonyme for Hairy.

Pilo'sity, *Pilo'sitas* (*Pilus* hair). Indicating the presence of Pubescence.

Pilosius'culus. Somewhat hairy.

Pil'ula (a little ball). Synonyme for *Galbulus*.

Pilus. Hair.

Pimpled. Covered with soft tubercular projections, or wart-like excrescences.

Pina'ceæ (from the genus Pinus). Synonyme for Coniferæ.

Pinna. A single leaflet in a pinnate leaf.

Pinnate, Pinna'ted, *Pinna'tus*, (feathered). When leaflets are arranged on opposite sides of a common petiole. When the arrangement is confined to pairs, it constitutes the "equally or pari-pinnate," (*a*). And when terminated by an odd leaflet, the "unequally or impari-pinnate," (*b*) fig. 125.

Pinnat'ifid, *Pinnati'fidus*, (*Pinnatus* feathered, *findo* to cleave). Where the lateral incisions of a simple leaf extend towards the axis, and approach the form termed Pinnate; fig. 126.

Pinnatiloba'te, *Pinnatiloba'tus*, (*Pinnatus* feathered, *lobus* a lobe). A variety of "pinnatifid;" where the incisions are not deep, or not very regularly disposed.

Pinnatisec'tus (*Pinnatus* feathered, *seco* to cut). A variety of "pinnatifid"; where the incisions are deep, extending to the axis.

PIN'NULA (a little feather). A leaflet of a compound leaf.

PIPERA'CEÆ (from the genus Piper). The Pepper tribe. A natural order of Dicotyledones.

PI'SIFORM, *PISIFOR'MIS* (*PISUM*, a pea, *FORMA*, shape). About the size and shape of a garden pea.

PISTIA'CEÆ (from the genus Pistia), The Duckweed tribe. A natural order of Monocotyledones; otherwise considered as a sub-order of Araceæ.

PIS'TIL, *PISTIL'LUM* (a pestle), Either synonymous with Carpel, when each portion of the innermost floral organ is free; or it is composed of the aggregation of the Carpels, when these unite and thus form a compound organ. It is essentially composed of the ovary (*o*), with its ovules; and the stigma or stigmata (*s*), with sometimes an intervening style (*s*); fig. 127.

PISTILLI'DIUM (diminutive of PISTILLUM). A young theca in Mosses, having the appearance of a pistil.

PITCHER. A peculiar form of leaf, in which the petiole expands into a hollow vessel, crowned by the limb, which, in some cases, assumes the appearance of a lid; fig. 128.

PIT'CHER-SHAPED. A tubular organ, bulging below and contracted towards the orifice.

PITH. A central column of cellular tissue, in the stems and branches of exogenous plants. The term is applied more generally to cellular parts; which are either called "piths," or said to be "pithy."

PIT'TED. Having numerous excavated spots or small depressions, on the surface.

PLACEN'TA, *PLACENTA'RIUM*. That part of the ovary which supports the ovules. The latter term is also extended, in some fruits, to the parts composed of the several placentæ, and now bearing the seed.

PLACEN'TA-SHAPED, *PLACENTIFO'RMIS*. A thickened circular disk, depressed in the middle both above and below.

PLACENTA'TION, *PLACENTA'TIO*. The arrangement of the seeds in the Pericarp. Also, the arrangement of the Cotyledons in the seed.

PLAIT'ED, Synonyme for Plicate.

PLANE, *PLA'NUS* (flat). Where a surface is perfectly level.

PLANIUS'CULUS. Where a surface is nearly, but not quite level.

Plantagina'ceæ, Plantagin'eæ (from the genus Plantago). The Rib-Grass Tribe. A natural order of Dicotyledones.

Plantula'tio. Synonyme for Germinatio.

Platana'ceæ, Plata'neæ (from the genus Platanus). The Plane tribe. A natural order of Dicotyledones.

Platycar'pus (πλατυς broad, καρπος fruit). Where the fruit of some species is remarkable for its breadth in comparison with fruit of a similar description. So also, for other compounds with πλατυς, as Platyphyllus, Platysiliquus, &c.

Plecolepi'dus πλεκω to bind, to fold; λεπις a scale). When the bracts forming the involucrum of Compositæ adhere together.

Ple'iophyllus (πλεῖος full, φυλλον a leaf). Where no buds, and consequently no branches, are developed in the axils of leaves, and the stems support these only. N. B. An error occurs in the definition given of ***Artiphyllus***, a term used in opposition to ***Pleiophyllus***. Its derivation is from ἄρτι a participle, which in composition indicates perfection; and not from ἄρθρον a joint.

Ple'nus (full). Synonyme for "double," where the stamens and pistils become petaloid. Also, where a stem is solid, in contradistinction to "hollow" or "fistular."

Pleuren'chyma (πλευρα a rib, εγχυω to diffuse). Woody tissue.

Pleuro-discus (πλευρα the side, δισκος a quoit). When some sort of appendage is attached to the sides of a disk.

Pleurogy'nus, Pleurogy'nius (πλευρα the side, γυνη a woman). Where a glandular or tubercular elevation rises close to, and is parallel with, the ovary.

Pleurogyra'tus (πλευρα the side, γυρος a circle). Where the "ring" on the "theca" of ferns is placed laterally.

Pleurorhi'zus (πλευρα the side, ῥίζα the root). When the radicle or the embryo is directed towards the hilum; as in orthotropous ovules. Synonyme for "accumbens," when applied to the embryo of Cruciferæ.

Plexe-oblas'tus (πλεγμα tissue, βλαςτος a shoot). When cotyledons rise above ground in germination, but do not assume the appearance of leaves.

Pli'ca (a plait or fold). Synonyme for Lamella, in some fungi. A diseased state in the developement of buds, which instead of forming true branches, become short twigs; and these produce others of the same sort, the whole forming an entangled mass.

PLICA'TE (*PLICATUS* knit together). Folded together in longitudinal plaits, regularly disposed, as in the vernation of some leaves; fig. 129.

129

PLICA'TILIS (*PLICO* to knit or fold together). Possessing the power or property of folding together; as the corollas of some flowers at distinct periods of the day.

PLICATI'VUS. Used synonymously with *PLICATUS*; but more especially applied as a synonyme for *CORRUGATUS*, where the longitudinal folds are irregularly wrinkled; as in some æstivations.

PLOCOCAR'PIUM (πλοκή a connection, καρπος a fruit). A fruit composed of follicles ranged round an axis.

PLUM'BEUS. Lead-coloured. Dull grey, with metallic lustre.

PLUMA'TUS (feathered). Synonyme for Pinnatus.

PLUMO'SE, *PLUMO'SUS* (full of feathers.) When hair is invested with branches, arranged like the beard on a feather.

PLU'MULE, *PLU'MULA* (a little feather). The portion of the embryo which developes in a contrary direction to the radicle. It is the first bud, or gemmule, of the young plant.

PLURIPARTI'TUS (*PLUS*, *PLURIS* more, *PARTITUS* proportionably divided). Where an organ is deeply divided into several nearly distinct portions.

PLUMBAGINA'CEÆ, PLUMBAGI'NEÆ (from the genus Plumbago). The Leadwort tribe. A natural order of Dicotyledones.

PNEU'MATO-CHYMI'FERUS (πνευμα air, χυμος juice, φερο to carry). A term applied to spiral vessels; and *PNEURATO-FERUS* has been restricted to the external membranous tube of such vessels.

POCULIFOR'MIS (*POCULUM* a cup, *FORMA* shape). Cup-shaped.

POD. Synonyme for Legume.

PODE'TIUM (πους a foot). A stalk-like elevation, simple or branched, rising from the thallus, and supporting the apothecium in some lichens. The term is also extended to the support of the fructification in Marchantia.

PODICIL'LUM (diminutive of Podetium). When the podetium is short.

PODOCAR'PUS (πους a foot, καρπος fruit). Where the ovary is seated on a gynophorus.

Podoce'phalus (πους a foot, κεφαλη the head). Where a head of flowers is elevated on a long peduncle.

Podogy'nium (πους a foot, γυνη a woman). Synonyme for Gynophorus.

Podogy'nicus, *Podo'gynus*. Synonyme for *Podocarpus*.

Podosper'mium, *Podosper'ma* (πους a foot, σπέρμα seed.

Podostema'ceæ, Podoste'meæ (from the genus Podostemon). A natural order of Dicotyledones.

Point'less. Without any sharpness at the termination; as in the case of parts furnished with an arista, seta, mucro, &c.

Point'letted. Synonyme for Apiculate.

Polake'nium (πολυς many, α not, χαινω to open). A fruit composed of closed carpels of the character possessed by akenia, invested by the calyx, and separable longitudinally. In the Umbelliferæ it is synonymous with Cremocarpium.

Polexos'tylus (πολυς many, εξω without, στυλος a style). Synonyme for Microbasis.

Pol'ished, *Poli'tus*. Perfectly smooth and glossy.

Pollachi'genus, (πολλακις frequently, γενναω to produce). A synonyme for Polycarpus.

Pollen (*Pollen* fine flour). Utricular grains, formed within the anther, either free and resembling dust, or variously agglutinated into waxy masses. The "granules" are very minute particles within the pollen-grains; and the "pollen-tube" is a membranous extension of a coat of the grain, developed when this is subjected to the influence of the stigma.

Pollen-mass. An agglutinated mass of pollen, peculiar to the state in which it occurs in some orders; as in Orchidaceæ, and Asclepiadeæ.

Pol'lex, *Pollica'ris* (a thumb's-breadth). About an inch long.

Pollina'ris, *Pollino'sus*. As if dusted with pollen.

Pollina'rium. Synonyme for Androcæum. Also used for an anther of Musci.

Polli'nicus. Composed of, or bearing some relation to, pollen.

Polyadel'phia (πολυς many, αδελφος a brother). An artificial class in the Linnean system, composed of plants whose stamens cohere into more than two distinct groups or bundles.

Polyadel'phous, *Polyadel'phicus*, *Polyadel'phus*. Having the stamens combined, as in Polyadelphia.

Polyan'dria (πολυς many, ανηρ a man). An artificial class

(also an order) in the Linnean system, composed of plants whose flowers have more than a dozen stamens, and are arranged hypogynously.

POLYAN'DROUS, *POLYAN'DER*, *POLYAN'DRICUS*, *POLYAN'DRUS*. A flower containing many stamens, arranged as in Polyandria. Also any flower with many stamens where the precise number is not named.

POLYAN'THEMUS, *POLYAN'THUS* (πολυς many, ανθος a flower). Bearing many flowers. Having many flowers aggregated. Investing many flowers.

POLYCAM'ARUS (πολυς many, καμαρα a vault). Synonyme for Polycarpicus; camara for carpella.

POLYCAR'PICUS POLYCAR'PUS (πολυς many, καρπος fruit). Where the carpels being distinct and numerous, each flower bears several fruit. Used synonymously with either perennial or arborescent plants, which, lasting many years, reproduce their fruit many times.

POLYCE'PHALUS (πολυς many, κεφαλη a head). When a common support is capped by many like parts.

POLYCHORION, *POLYCHO'RIS*, *POLYCHORIONI'DES* (πολυς many, χοριον fœtal membrane). Synonyme for Etærio.

POLYCLA'DIA (πολυς many, κλαδος a branch). Where there is a supernumerary development of leaves and branches; as in the diseased state called Plica.

POLYCLO'NUS (πολυς many, κλων a branch). Where a stem is much branched.

POLYCOCCUS (πολυς many, κοκκος a seed). A fruit composed of many Cocca.

POLYCOTYLE'DONOUS, *POLYCOTYLEDO'NEUS*, *POLYCOTYLE'DONUS* (πολυς many, κοτυληδων (see Cotyledon). With more than two cotyledons. Also used synonymously with Dicotyledonous, in opposition to Monocotyledonous.

POLYFLO'ROUS (πολυς many, flos a flower). Synonyme for Multifloral.

POLYGALA'CEÆ, POLYGA'LEÆ (from the genus Polygala). The Milk-wort tribe. A natural order of Dicotyledones.

POLYGA'MIA (πολυς many, γαμος a marriage). An artificial class of the Linnean system, including plants which bear three descriptions of flowers; viz., hermaphrodite, male, and female.

POLYG'AMOUS, *POLYG'AMUS* (πολυς many, γαμος a marriage). Plants with flowers formed as in Polygamia.

POLYGONA'CEÆ, POLYGO'NEÆ (from the genus Polygonum). The Buck-wheat tribe. A natural order of Dicotyledones.

POLYGONA'TUS (*πολυς* many, *γονυ* a knot). Where the stem has many knots.

POLYGO'NUS (*πολυς* many, *γωνια* angle). Synonyme for Multangular.

POLYGY'NIA (*πολυς* many, *γυνη* a woman). An artificial order under the Linnean system, containing plants which have many pistils; or, at least many distinct styles, if the ovary is compound.

POLY'GYNOUS, *POLYGY'NICUS*, *POLYGY'NUS* (*πολυς* many, *γυνη* a woman). Having many distinct pistils; or an ovary with many distinct styles.

POLYGY'RUS (*πολυς* many, *γυρος* a circle). In several whorls or circles.

POLYLE'PIDUS (*πολυς* many, *λεπις* a scale). Furnished with many scales.

POLY'MERUS (*πολυς* many, *μερος* a part). Composed of many like parts.

POLYMOR'PHOUS, *POLYMOR'PHUS* (*πολυς* many, *μορφη* form). Where any part, or an entire species, is subject to considerable diversity of form.

POLYNE'URIS (*πολυς* many, *νευρον* a nerve). Where the nerves of a leaf, but especially the secondary, are numerous.

POLYOVULA'TUS (*πολυς* many, *OVUM* an egg). Furnished with many ovules.

POLYPE'TALOUS, *POLYPET'ALUS* (*πολυς* many, *πεταλον* a leaf). Where the petals of a corolla form no cohesion; but are perfectly distinct.

POLYPHO'RE, *POLY'PHORUM* (*πολυς* many, *φερω* to bear). A receptacle with many distinct carpels.

POLYPHYL'LUS (*πολυς* many, *φυλλον* a leaf). Any foliaceous assemblage composed of many subordinate pieces.

POLYPODIA'CEÆ (from the genus Polypodium). Either a sub-order of Filices; or a distinct order, when the entire group of Ferns is considered to be thus sub-divisable.

POLYRHI'ZUS (*πολυς* many, *ριζα* a root). Possessing numerous rootlets. Where flowering parasites, attached to roots, have many distinct rootlets, independently of those by which their attachment is effected.

POLYSE'CUS (*πολυς* many, *σηκος* a stalk). Synonyme for Etærio.

Polyse'palous, *Polyse'palus* (πολυς many, *sepalum* a word coined for sepal). Where the sepals of a calyx form no cohesion.

Polysper'mus (πολυς many, σπερμα seed). Where a pericarp contains numerous seeds.

Polyspo'rus (πολυς many, σπορα seed). When the theca &c., of acotyledonous plants contains many spores.

Polysta'chyus (πολυς many, σταχυς a spike). Where many spikes are combined in the inflorescence.

Polystem'onous, *Polystem'onus* (πολυς many, στημων a stamen). Possessing many more stamens than petals.

Polys'tigmus (πολυς many, στιγμα the stigma). Where a flower has many carpels, each originating a stigma.

Polys'tomus (πολυς many, στομα a mouth). Where root parasites are attached by means of a succor at the extremity of each of many rootlets.

Polys'tylus (πολυς many, στυλος a style). Where a compound ovary has several distinct styles.

Polythe'leus (πολυς many, θηλη a nipple). A flower containing several distinct ovaries.

Poly'tomous, *Poly'tomus* (πολυς many, τεμνω to cut). Where the limb of a leaf is distinctly subdivided into many subordinate parts, but these are not jointed to the petiole, and therefore not true leaflets, and the leaf itself not compound.

Pome, *Po'mum* (an apple). A fleshy multilocular fruit, matured from an inferior ovary.

Pomeridia'nus (in the afternoon). Applied to flowers which expand after noon.

Pomif'erus (*pomum* an apple, *fero* to bear). Bearing fruit, or even excrescences, shaped like an apple.

Pomifor'mis (*pomum* an apple, *forma* shape). Approaching the shape of an apple.

Pomol'ogy, *Pomolo'gia* (*pomum* a general name for fruit, λογος a discourse). A treatise on fruits.

Pontedera'ceæ, Pontede'ræ, Pontederia ceæ (from the genus Pontederia). A natural order of Monocotyledones.

Poren'chyma (πορος a pore, εγχυμος succulent). Cellular tissue, elongated and apparently perforated by pores.

Poro'sus (πορος a pore). Where the tissue is, or appears to be, full of small holes.

Porphy'reus (πορφυρεος purple). Synonyme for Purpureus,

Portulaca'ceæ, Portula'ceæ (from the genus Portulaca). The Purslane tribe. A natural order of Dicotyledones.

Por'ulus (diminutive from *Porus*).

Po'rus and pl. ***Po'ri*** (πορος a pore). A minute superficial hole. Synonyme for Stoma, and for Ascus of some fungi.

Posti'cus (behind). Synonyme for Extrorsus.

Potalia'ceæ (from the genus Potalia). A natural order of Dicotyledones.

Pota'meæ (ποταμος a river). Synonyme for Fluviales.

Pouch-shaped. Resembling a little bag, generally double, or two-celled. Also applied to the twin tuberous roots of some Orchideæ. Same as Scrotiform.

Powdery. When a surface is coated by a fine powder, as the bloom on Plums, said to be of waxy nature.

Præcox (early). Appearing or flowering earlier than other allied species.

Præflora'tion, *Præflora'tio* (*præfloreo* to blossom before the time). Synonyme for "Æstivation."

Præfolia'tion, *Præfolia'tio* (*præ* before, *folium* a leaf). Synonyme for "Vernation."

Præmorse, *Præmor'sus* (bitten off). Ending abruptly; as where roots, or more frequently rhizomata (called roots), have decayed at the end; fig. 130, *a*. Also where the truncate termination of any foliaceous lamina appears jagged; fig. 130, *b*.

Præus'tus (burnt at the point). Of a brown tint, as though it had been produced by burning.

Pra'sinus (leek-green). Pure green with a slight admixture of grey.

Praten'sis (belonging to a meadow). When the botanical station of a plant is generally that of a meadow.

Pre'cius (ripening before others). Synonyme for Præcox.

Prickle. A more or less conical elevation of the substance of the bark, corky within, but hard and sharp-pointed.

Prickly. Bearing prickles on the stems or branches.

Pri'mary, *Prima'rius*. The part which is first developed. The principal parts in the subdivisions of a compound arrangement. As the main petioles of a compound leaf; the pedicels of a compound umbel which support the partial umbels.

Primige'nius (natural, original). Synonyme for *Primitivus*.

PRI'MINE, *PRI'MINA* (PRIMUS first). The outermost and last developed integument to the nucleus of the ovu.e.

PRIM'ITIVE, *PRIMITI'VUS* (the first or earliest). The first parts developed. Applied to specific types; in opposition to forms resulting from hybridisation.

PRIMOR'DIAL, *PRIMORDIA'LIS* (PRIMORDIUM the beginning or origin). The earliest formed of any set of organs; or the principal parts of some of them.

PRIMOR'DIAL-UTRICLE. The layer or internal coating first formed in, and co-extensive with, the cell, during the perfecting of the cellular tissue.

PRIMULA'CEÆ (from the genus Primula). The Primrose tribe. A natural order of Dicotyledones.

PRISMAT'ICAL, PRISM-SHAPED, *PRISMAT'ICUS* (πρισμα a prism). Approaching the form of a prism; where the surface of either solids or tubes presents angles, disposed longitudinally.

PRISMEN'CHYMA (πρισμα a prism, εγχυμος succulent). Where the vesicles of the cellular tissue are prismatic,

PROBOCID'EUS (PROBOSCIS the trunk of the elephant). Beaked.

PROCE'RUS (lofty). Synonyme for Elatus.

PRO'CESS, *PROCES'SUS* (progress). An extension or projection from a surface.

PROCUM'BENT, *PROCUM'BENS* (bending downwards). Lying upon or trailing along the ground.

PRODUC'TUM (PRODUCTUS prolonged). Synonyme for Calcar.

PROEM'BRYO (*PRO* for, instead of, *EMBRYO* the embryo). The portion of the spore in some Acotyledones which assumes a foliaceous character in developing.

PROEM'INENS (*PRO* in comparison of, *EMINENS* rising up). When a part is unusually extended or stretches beyond another, which more frequently surpasses it.

PROJECT'URA (the jutting out of a building). A small longitudinal projection on some stems where the leaf originates.

PROLIF'EROUS, *PRO'LIFER*, *PROLI'FERUS* (PROLES a race, FERO to bear). An unusual development of supernumery parts, of the same or of a different description from those on which they are developed. As where the flower buds become viviparous; where the leaf produces gems, &c.

PROLI'GERUS (PROLES, offspring; GERO, to bear). Applied to a portion of the apothecia of lichens, in which the sporules are generated.

Prom'inent, *Prom'inens* (standing out). Projecting beyond some neighbouring part.

Pro'nus (face downwards). Lying flat upon the ground, or other support. Applied also to the under-surface of a horizontal leaf.

Propac'ulum (Propago, a shoot). An offset.

Propa'go (a shoot). A layer. An axillary bud.

Propag'ulum. Synonyme for Propaculum. A granular reproductive body, many of which, collected together, form the soredia of lichens.

Pro'physis, *Pros'physis* (προφυω to beget before). Synonyme for Adductor.

Pro'prius (peculiar). Partial.

Proscol'la (προσκολλαω to glue to). A viscid gland on the rostellum of an Orchis.

Prosem'bryum (προς near, about, εμβρυον the embryo). Synonyme for Perispermium.

Prosen'chyma (προς near, εγχυμος succulent). Masses of cellular tissue composed of utricles which are more or less fusiform; the tapering extremities overlapping or interlacing.

Pros'physis *Pros'physus* (προς near, φυσις growth). Synonyme for Adductor.

Pros'trate, *Prostra'tus* See Procumbent.

Pros'typus (προστυπος embossed). Synonyme for Raphé.

Protea'ceæ (from the genus Protea). A natural order of Dicotyledones.

Proteran'thous, *Proteran'thus* (προτερος before, ανθος a flower). Where the flower-buds expand before the leaf-buds.

Protophyl'lum (πρωτος first, φυλλον leaf). A seminal leaf. Or, more especially, restricted to the first leaf of an Acotyledonous plant.

Protophyto'logy (πρωτος first, φυτον plant, λογος discourse). Fossil botany.

Protophy'tum (πρωτος first, φυτον plant). Has been applied synonymously with Alga and Lichen.

Protoplas'ma (πρωτος first, πλασμα a formed work). A coating deposited on the inside of the cells of the cellular tissue; and considered to be a nitrogenized compound.

Protos'trophis (πρωτος first, τροφη a good). A spiral vessel, separating from the main bundle that enters the leaf, and forming part of the primary veins.

Protothal'lus (πρωτος first, θαλλος a frond). The first part formed, or substratum to the thallus of lichens.

Prui'na (hoar-frost). Powdery secretion on the surface of some plants; also on certain fruits.

Pruina'tus, *Pruino'sus* (frosty). Frosted. Powdery.

Prunifor'mis (*prunum* a plum, *forma* shape). Approaching the form of a plum.

Pru'num (a plum). Synonyme for Drupa.

Pru'riens (itching). Producing an itching sensation.

Pseu'do-bulb. Swollen internodes of many of the Orchideæ, resembling bulbs.

Pseudo-car'pus (ψευδης false, καρπος fruit). Synonyme for Galbulus.

Pseudo-costa'tus (ψευδης false, *costatus* ribbed). Where the outer veins of a leaf combine, and form a line parallel to the margin.

Pseudo-cotyl'edones (ψευδης false, κοτυληδων cotyledon). A group of Acotyledones, including those Orders where the the Proembryo, in developing, assumes the appearance of a cotyledon.

Pseudo-gyra'tus (ψευδης false, γυρος a circle). Where the annulus of a fern is seated on the summit of the theca.

Pseudo-hyme'nium (ψευδης false). A covering of the spores of Algæ, resembling the hymenium of fungi.

Pseudo-monocotyle'dones (ψευδης false). When the cotyledons of a dicotyledonous plant cohere, and thus appear as if they were only one.

Pseudo-parasit'icus (ψευδης false, παρασιτικος parasitical). Synonyme for Epiphytic. Also, deriving nourishment from dead, not living, organic tissue.

Pseudo-perid'ium. Resembling a peridium.

Pseudo-peristo'mium. The external peristome, where it is early obliterated.

Pseudo-perithe'cium. Resembling a perithecium.

Pseudo-po'dium (ψευσης false, πους a foot). A leafless dilated branch, on which the sessile theca of some mosses is developed.

Pseudo-pyren'ium. Synonyme for Perithecium in certain fungi.

Pseudo-sper'mus (ψευδης false, σπερμα seed). Has been used to express the nut-like carpels of the Labiatæ and Boragineæ.

Pseudo-ste'reus (ψευδης false, στερεος solid). Becoming partially coherent, or grafted together.

Pseudo-stro'ma. Synonyme for Perithecium in certain fungi.

Pseudo-thal'lus. The axis of densely crowded forms of inflorescence.

Psilosta'chyus (ψιλος thin, σταχυς a spike). Where the inflorescence is in very slender spikes.

Psydomor'phytus (ψευδης false, μορφη form, φυτον plant). Where a capitate inflorescence affects the form of the Capitulum in Compositæ.

Pteri'dies, Pterid'ium, (πτερον a wing). Synonyme for Samara.

Pterido'graphia, Pterigraph'ia (πτερις a fern, γραφω to write). A treatise on Ferns.

Pteri'gynus (πτερὸν a wing, γυνη a woman). Synonyme for Pterospermus.

Pterocar'pus (πτερον a wing, καρπος fruit). Where a fruit is winged.

Pterocau'lis (πτερον a wing, καυλος a stem). Where a stem is winged.

Ptero'dia, Ptero'dium. Synonyme for Pteridies.

Pterog'onus (πτερον a wing, γωνια an angle). Where an angle is winged.

Pteroi'deus (πτερον a wing, ειδος resemblance). An elevated extension of the surface assuming a wing-like appearance.

Ptero'podus (πτερον a wing, πους a foot). Where the petiole is winged.

Pterosper'mus (πτερον a wing, σπερμα a seed). Where a seed is winged.

Ptery'gium (πτερον a wing). A wing.

Pteryg'opus (πτερον a wing, πους a foot). Where the peduncle is winged.

Pterygosper'mus (πτερον a wing, σπερμα a seed). Synonyme for Pterospermus.

Pty'chodes (πτυξ a fold). Synonyme for Protoplasma.

Pube'ns. Synonyme for Pubescens.

Pu'berty, *Puber'tas.* The period when a plant first begins to produce flowers.

Pu'ber (full age). The period of maturation in fruit.

Pu'bescence (*pubes, pubescentia* down). Elevated extensions of the cellular tissue of the epidermis, assuming the character of hair, scale, gland, &c.

PUBES'CENT, (*PUBENS, PUBESCENS*, downy). Furnished with pubescence.

PUBIG'ERUS (*PUBES* down, *GERO* to bear). Synonyme for Pubescens.

PUGION'IFORMIS (*PUGIO* a dagger, *FORMA* shape). Shaped like a dagger.

PULLEY-SHAPED. A cylinder, gradually contracting towards the middle, with the extremities hemispherical.

PUL'LUS (πελλος black). Synonyme for Coracinus.

PULP (*PULPA*). Soft and juicy tissue.

PULPY, *PULPO'SUS*. Of the consistence of pulp.

PULVERA'CEUS PULVE'RIUS (*PULVIS* dust). Powdery.

PULVERULEN'TUS (dusted). Coated with powdery grains.

PULVINA'TUS (made like a cushion), *PULVINIFOR'MIS* (*PULVINUS* a cushion, *FORMA* shape). Assuming the appearance of a cushion or pillow.

PULVIN'ULA, PULVIN'ULUS (diminutive of Pulvinus). Simple or branched excrescences originating on the upper surface of the thallus of some Lichens.

PULVI'NUS (a cushion). An enlargement, like a swelling, on the stem immediately below the leaf. Also an enlargement of the base in some petioles. Fig. 131.

131

PUL'VIS (dust). Any light powder excreted on the surface.

PULVIS'CULUS (*PULVIS* dust). The powder contained in the spore-cases of some fungi.

PU'MILUS (little). Dwarfish, in comparison with allied species.

PUN'CTATE, *PUNCTA'TUS, PUNCTICULA'TUS, PUNCTICULO'SUS*, (*PUNCTUM* a point). Synonyme for "Dotted."

PUNCTIFOR'MIS (*PUNCTUM* a point, *FORMA* shape). In the form of a small pointed projection; or nearly redueed to a mere point.

PUN'GENT, *PUN'GENS*, (pricking). Very hard and sharp-pointed.

PUNI'CEUS (scarlet). Pure red.

PUR'PLE, *PURPURA'RIUS, PURPURA'TUS, PURPU'REUS* (of a purple colour). Blue and red.

PURPURAS'CENS. Inclining to a purple colour.

Purse-shaped. Synonyme for Pouch-shaped.

Pusil'lus (weak and small). Diminutive, with respect to allied species.

Pustula'tus, *Pustulo'sus* (*pustula* a blister). Having convex elevations like blisters.

Puta'men (shell of a nut). The endocarp, when it becomes hard and bone-like; as in stone fruits.

Putamina'ceus. Bony.

Pycnoceph'alus (πυκνος dense, κεφαλος head). Where the flowers are densely crowded in the inflorescence.

Pygmæ'us (dwarfish). Synonyme for Pumilus.

Pyracan'thus (πυρ fire, ακανθα a spine). With yellow spines.

Pyram'idal, *Pyramid'alis* (*pyramis* a pyramid). Either angular and tapering upwards, as a pyramid; or used synonimously with Conical.

Pyre'ne, *Pyre'na* (*pyren* kernel, or stone of fruit). Synonyme for *Putamen*. Also, synonyme for Nucule.

Pyrena'rium, *Pyrid'ium*. Synonymes for Pomum; more especially when the endocrap is bony.

Pyrif'erus *Pyrifor'mis* (*pyrus* a pear, *fero* to bear. *forma* shape). Pear-shaped.

Pyxida'tus (like a box with a lid). Furnished with, or formed like, a Pyxidium.

Pyxid'ium, *Pyxid'ula*, *Pyx'is* (a box). A capsule with transverse dehiscence, which separates it into two parts; the lid (*operculum*) and the urn (*amphora*); Fig. 132. Also, synonyme for the Theca of Mosses.

132

Quadran'gular *Quadrangula'ris* *Quadran'gulus* (four-cornered), *Quadrangula'tus*. Approximating to the form of a quadrangular prism.

Quadricru'ris (*quadrans* a fourth part, *crus* a leg). Where the supports are four.

Quadricotyledon'eus. A Dicotyledonous plant with four cotyledons.

Quadridi'gitate, *Quadridigita'tus* (*quadrans* a fourth part, *digitatus* having fingers). Digitate in four divisions.

Quadrifa'rius (*quadrifariam* four ways). Arranged in four rows.

Quadrifo'liate, *Quadrifolia'tus*. Synonyme for Quadridigitate.

QUADRI'FIDUS (cleft in four parts). Where the sub-division of an organ into four parts extends to about the middle, or half way down.

QUADRIHILA'TUS (*HILUM* a spot). Having four apertures.

QUADRI'JUGUS (*JUGUM* a yoke). In four pairs.

QUADRINA'TUS, *QUADRI'NUS* (belonging to four). Where four leaflets meet at the extremity of a petiole in a digitate arrangement.

QUADRIPARTI'TUS (*PARTITUS* proportionably divided). See Parted. The four sub-divisions extend deeper than in Quadrifidus.

QUAR'TINE, *QUARTI'NA* (*QUARTUS* the fourth). A lamina, resembling a distinct integument, which occasionally occurs within the tercine or nucleus of ovules.

QUASIRADIA'TUS (*QUASI* as if, *RADIATUS* with rays). Where the florets of the ray, in the capitula of Compositæ, are inconspicuous. Also applied to the Periclinium when it appears to be only slightly radiate.

QUATER'NATE, *QUATERNA'TUS* (*QUATERNI* in fours). When verticillate appendages are arranged by fours.

QUERCI'NÆ (from the genus Quercus). Synonyme for Cupuliferæ.

QUI'NATE, *QUINA'TUS* (*QUINUS* five). Where five similar parts are arranged together; as five leaflets in a digitate leaf.

QUINCUN'CIAL, *QUINCUNCIA'LIS* (in the order of the Quincunx). When the parts of a floral whorl, in æstivation, are five, and so disposed that two are exterior, one or two wholly interior, and the other two, or one, partially imbricate. Fig. 133.

QUINQUEFA'RIUS. Disposed longitudinally in five rows.

QUINQUENER'VED, *QUINQUENER'VIS*, *QUINQUENER'VIUS* (*QUINQUE* five, *NERVA* a nerve). When the primary nerves of a leaf, four in number, branch off from the base of the limb, so that (including the mid-nerve) it becomes furnished with five ribs. Fig. 134.

QUIN'TINE, *QUINTI'NA* (*QUINTUS* the fifth). A lamina resembling a distinct integument, which occasionally invests the embryo, within the Quartine.

QUIN'TUPLE-NERVED, *QUINTU'PLI-NERVIS*. Where four distinctly marked primary nerves of a leaf are given off from the mid-rib, but do not meet at the base as in Quinque-nerved.

QUINTU'PLED, *QUINTU'PLEX* (five fold). Where the arrangement is a multiple of five.

RACE. A variety of any species of which the individuals, for the most part, retain a marked peculiarity of character when raised from seed.

RACE'ME, *RA'CEMUS* (a bunch). A form of inflorescence, where the flowers are furnished with pedicels arranged at intervals upon a common axis. Fig. 135.

135

RACEMIF'ERUS, *RACEMIFLO'RUS* (*RACEMUS* a cluster, *FERO* to bear, *FLOS* a flower), Synonyme for Racemosus.

RACEMIFOR'MIS (*RACEMUS* a cluster, *FORMA* shape). Where a Thyrse assumes the appearance of a raceme, from the peduncle bearing only one or extremely few flowers.

RACEMO'SE, *RACEMO'SUS* (full of clusters). The inflorescence in racemes.

RACEMULO'SUS (diminutive of *RACEMOSUS*). Inflorescence in very small racemes.

RACHIMOR'PHUS (ραχις the back bone, μορφη form). The floriferous axis of spiked grasses.

RA'CHIS (ραχις the back bone). The axis of several kinds of inflorescence. The stalk or petiole to the fronds of ferns.

RACHI'TIS (ραχιτις in the back bone). A disease producing abortion in the fruit or seed.

RA'DIAL, *RADIA'LIS* (*RADIUS* a sun beam). Belonging to the ray, in Compositæ, &c.

RA'DIANT, RA'DIATE, RADIA'TED, RADIA'TING, *RA DIANS*, *RADIA'TUS* (*RADIUS* a sun beam). Arranged like rays, or the extreme portion of rays, spreading from a common centre.

RADIATIFO'RMIS (*RADIUS* a sun beam, *FORMA* shape). Where the florets of a liguliflоral capitulum, in Compositæ, increase gradually in length, from the centre towards the circumference, with the corolla extending outwards.

RADIA'TIM (*RADIUS* a sun beam). Arranged in the manner explained under Radiant.

RAD'ICAL, *RADICA'LIS* (*RADIX* the root). Proceeding from a point close to the summit or crown of the root.

RADI'CANS (*RADICOR* to take root). Rooting.

RADICA'TION, *RADICA'TIO*, *RADICELLA'TIO* (*RADIX* a root). The general disposition and arrangement of the roots of a plant.

RADICA'TED, *RADICA'TUS* (*RADICOR* to take root). Possessing roots; or, more especially, furnished with a decidedly marked tap root.

RADICEL'LA (diminutive for *RADIX*). Synonyme for Radicula.

RADICELLA'RIS (from *RADICELLA*). With very small roots. Or, bearing reference to the Radicle.

RADICI'COLUS (*RADIX* a root, *COLO* to inhabit). Parasitic on the roots of plants.

RADI'CIFLORUS (*RADIX* a root, *FLOS* a flower). Where the flower is seated immediately above the crown of the root; or where it rises from an under-ground rhizoma.

RADICIFOR'MIS, *RADICI'NUS* (*RADIX* a root, *FORMA* shape). Presenting the general appearance of a root; or having the usual consistency of a root.

RAD'ICLE, *RADI'CULA* (diminutive for Radix). The rudimentary state of the root in the embryo. Sometimes applied to small roots, or to the fibres about the tap-root.

RADICULIFOR'MIS (*RADICULA* a radicle, *FORMA* shape). Having the appearance of roots, but serving only as means of support, and not otherwise performing the functions of true roots.

RADICULO'DA, *RADICULO'DIUM* (*RADICULA* a radicle, εἶδος resemblance). A synonyme for Radicula; or rather for the apex of it, where it receives the name of Blastus.

RA'DIUS (a sun beam). The Ray. Also a partial peduncle in Umbelliferæ.

RA'DIUS-MEDUL'LARIS (*RADIUS* a sun beam, *MEDULLA* the pith). A medullary-ray.

RA'DIX. The Root.

RAFFLESIA'CEÆ (after the genus Rafflesia). An order of flowering rhizanthous parasites.

RA'MAL, *RAMEA'LIS* (*RAMEUS* belonging to a branch). Either originating from a branch, or merely growing on a branch.

RAMAS'TRUM (*RAMUS* a branch). A partial petiole.

RAMEA'RIUS (*RAMEUS* belonging to a branch). Restricted to aerial roots, which originate from branches.

RAMENTA'CEOUS, *RAMENTA'CEUS*. Bearing ramenta.

RAMEN'TUM (a shaving). A thin membranous scale-like lamina of cellular tissue, on the surface of some plants.

RA'MEOUS, *RA'MEUS*. Synonyme for Ramal.

RAMIF'ERUS, *RAMIFICA'TUS* (*RAMUS* a branch, *FERO* to bear). Synonyme for Ramosus.

Ramiflo'rus (*ramus* a branch, *flos* a flower). Where the inflorescence occurs on the branches.

Ramifor'mis (*ramus* a branch, *forma* shape). Resembling a branch in form.

Rami'parus (*ramus* a branch, *paro* to produce). Synonyme for Ramosus.

Ramo'sus (full of branches). Producing branches; or very much branched.

Ramulo'sus (*ramulus*). Bearing Ramuli.

Ra'mulus (a little branch) *Ramun'culus*. The ultimate sub-division in branching. Also caulinar appendages assuming the form of branches.

Ra'mus (a branch). Any sub-divisions of the stem, originating in the development of a caulinar leaf-bud.

Ramus'culum, *Ramus'culus* (a little branch). Synonymes for Ramunculus. Ramusculi is applied to the mycelium of some Fungi.

Ranuncula'ceæ (from the genus Ranunculus). The Crowfoot tribe. A natural order of Dicotyledones.

Rapa'ceus (*rapa* in allusion to the root of a Radish). Synonyme for Fusiformis.

Raphe', *Rapha* (ραφη a seam). A fibro-vascular chord running from the placenta to the nucleus, through the chalaza. See fig. 47.

Raphi'de, *Ra'phida*, *Ra'phis* (ραφις a needle). A minute, frequently acicular, crystal of some insoluble salt formed in the interior of plants.

Ra'rus (thinly set). Where particular organs are not crowded; or fewer in number than is usual in allied species.

Ra'ven-black. See Coracinus.

Ray. The outer florets in a capitulum in Compositæ. The outer flowers, when differently formed from the inner, in umbels.

Reaumurea'ceæ (from the genus Reaumuria). A natural order of Dicotyledones.

Recep'tacle, *Recepta'culum* (a store-house). A support to one or more organs of the same description. It has been used synonymously with Amphanthium, Clinanthium, Torus, Placenta. It is also applied to various forms of support to the fructification of cryptogamous plants. Likewise to the chambers or cysts in which various secretions are deposited.

Receptacula'ris. Where there is attachment to some form of receptacle.

Re'cess (*recessus*). Synonyme for Sinus.

Recli'nate, Recli'ning, *Reclina'tus* (lying all along). So far bent from a perpendicular direction that the upper end becomes directed towards the ground. Also implies that one part is pressed down upon another.

Reclu'sus (disclosed). Has been strangely used synonymously with Inclusus.

Recondi'tus. Hidden.

Rectiflo'rus (*rectus* straight, *flos* a flower). Where the axes of the florets, in some Compositæ, are parallel to the main axis of the inflorescence.

Rectiner'vis, Rectiner'vius (*rectus* straight, *nervus* a nerve). Synonyme for Parallelinervis.

Rective'nius (*rectus* straight, *vena* a vein). Synonyme for Rectinervis.

Recur'ved, *Recurva'tus Recur'vus* (crooked). Synonyme for Curved; but especially when the bending is in a backward direction.

Recuti'tus (skinned). Having the appearance of being divested of epidermis.

Redu'plicate, Redu'plicative, *Reduplica'tus, Reduplicati'vus* (*re* from, *retro* back, *plico* to fold). Valvate, with the edges reflexed.

Reflex', Reflex'ed, *Reflex'us* (turned back). Where the apex is so far bent back as to approach the base.

Refrac'tus (broken). Where a part is so suddenly reflex, as to appear broken at the point where curvature takes place.

Reg'ma (ρηγμα rupture). Synonyme for Coccum.

Regres'sus (returning). Synonyme for Reflexus. Where a floral organ assumes the character of another which belongs to the whorl preceding that in position to which itself belongs.

Reg'ular, *Regula'ris* (according to rule). Uniformity in structure or condition. Where subordinate parts of the same kind closely resemble each other, and are symmetrically arranged.

Regulariflo'rus (*regularis* regular, *flos* flower). When the capitulum, or the disk only, of some of the Compositæ, consists of floscular florets.

REGULA'RIFORMIS (*REGULARIS* regular, *FORMA* shape). Closely approximating to regularity of structure.

REL'IQUIÆ (leavings). Appendages which wither without falling off. Synonyme for Induviæ.

REMO'TUS (removed). Synonyme for Rarus.

RE'NIFORM, *RENA'RIUS*, *RENIFOR'MIS* (*REN* the kidneys, *FORMA* shape). Resembling the section, taken longitudinally through a kidney; fig. 136.

REPAND', *REPAN'DUS* (bowed). When the margin is uneven, bespeaking a tendency to become sinuate.

RE'PENS (creeping). Lying flat upon the ground, and emitting roots along the under surface.

REPLICA'TE, REPLICA'TIVE, *REPLICATI'VUS*, *REPLICA'TUS* (unfolded). Doubled down, so that the upper part comes in contact with the lower. In estivation the fold is inwards, in vernation backwards.

RE'PLUM (a door cheek). The presistent portion of some pericarps, after the valves have fallen away.

REPRODUCTIVE ORGANS. The parts of a plant immediately concerned in the formation of seeds, sporules, &c.

REP'TANS (creeping along). Synonyme for Repens.

RESEDA'CEÆ (from the genus Reseda). The Mignionette tribe. A natural order of Dicotyledones.

RESINIF'ERUS (*RESINA* resin, *FERO* to bear). Secreting resin.

RESPIRA'TION, *RESPIRA'TIO* (breathing). A function of the leaf, and other parts, furnished with stomata, by which carbonic acid is decomposed, and carbon assimilated into an organic compound.

RES'TANS (remaining). Synonyme for Persistens.

RESTIA'CEÆ (from the genus Restio). A natural order of Monocotyledones.

RESTIB'ILIS (producing every year). Synonyme for Perennis.

RESU'PINATE, *RESUPINA'TUS* (lying on the back). So turned or twisted that the parts naturally the undermost become the uppermost, and vice versâ.

RE'TE (a net). That portion of the surface of the receptacle in compositæ which surrounds the *AREOLÆ*, in which the ovaries are seated. Filamentous web extending in some agarics from the edge of the pileus to the stipes.

RETICULA'TO-VENO'SUS (*RETICULATUS* made like a net, *VENA* a vein). Synonyme for Retinervis.

RETICULATE, *RETICULA'TUS* (made like a net). Resembling net-work.

RETIC'ULUM (a little net). The debris of crossed fibres about the base of the petioles in Palms.

RETI'FERUS, *RETIFOR'MIS*, (*RETE* a net, *FERO* to bear, *FORMA* shape). Synonymes for Reticulatus.

RETINACULA'TUS. Hooked.

RETINAC'ULUM (a stay or hold-fast). A viscid gland connected with the stigma in Orchideæ and Asclepiadeæ, which retains the pollen mass.

RETINER'VIS, *RETINER'VIUS* (*RETE* a net, *NERVA* a nerve). Where the nerves of leaves, &c., are reticulate.

RETRAC'TUS (drawn back). Where cotyledons are so far prolonged at their base as to completely conceal the radicle.

RETROCURVA'TUS, *RETROCUR'VUS*, (*RETRO* backward, *CURVATUS* bent). Synonymes for Recurvus.

RETRO'FLEXUS (*RETRO* backward, *FLEXUS* bent). Synonyme for Reflexus.

RETROFRAC'TUS (*RETRO* backward, *FRACTUS* broken). Synonyme for Refractus.

RETROR'SUM, *RETROR'SUS* (backward). Used synonymously with Recurvus and Reflexus.

RETROVER'SUS (*RETRO* backward, *VERSUS* turned). Synonyme for Inversus.

RETUSE, *RETU'SUS* (Blunted). Having a slight depression or sinus at the apex. Fig. 137.

REVOLU'BILIS (capable of being rolled back). Synonyme for Revolutivus.

RE'VOLUTE, *REVOLU'TUS* (turned back). Rolled backwards from the extremity upon the under-side or surface.

REVOLUTI'VUS (*REVOLUTUS* turned back). Rolled backwards from the margins upon the under surface.

RHAB'DUS (ῥαβδος a rod). The stipes of certain Fungi.

RHAMNA'CEÆ, RHAM'NEÆ, RHAM'NI (from the genus Rhamnus). The Buck-thorn tribe. A natural order of Dicotyledones.

RHEG'MA. See Regma.

RHINANTHA'CEÆ (from the genus Rhinanthus). Synonyme for Scrophulariaceæ.

RHIZAN'THEÆ (ῥιζα a root, ανθος a flower). A group containing a few Orders of flowering parasites, which mostly attach themselves to the roots of plants, and whose seeds exhibit a very imperfectly developed embryo.

Rhizan'thus (ῥίζα a root, ανθος a flower). Synonyme for Radicalis.

Rhi'zina (ριζα a root). Distinguishes the peculiar roots of Mosses and Lichens.

Rhizoblas'tus (ῥῖζα a root, βλαστος a germ). An embryo provided with an incipient root.

Rhizobola'ceæ, Rhizoboleæ (from Rhizobolus, a synonyme for the genus Caryocar). A natural order of Dicotyledones.

Rhizocar'pous, *Rhizocarpia'nus Rhizocar'picus* (ριζα a root, καρπος fruit). An herbaceous perennial.

Rhizo'genum (ῥίξα a root, γενναω to produce). The dilated base of the frond, in certain Algæ, from which proceed root-like appendages by which it adheres to its support.

Rhizoi'deus (ριζα a root, ειδος resemblance). Resemblıng a root in general appearance.

Rhizo'ma (ριζα a root). A prostrate or subterranean stem, from which roots are emitted, and scaly leaves or branches given off at the knots. A synonyme for Caudex. A synonyme for Radicula.

Rhizoma'ticus. Having the character of a Rhizoma.

Rhizomor'phus (ριζα a root, μορφη form). Assuming the appearance of a root.

Rhizo'philus (ριζα a root, φῖλος a friend). Growing attached to roots.

Rhizophora'ceæ, Rhizopho'reæ (from the genus Rhizophora). The Mangrove tribe. A natural order of Dicotyledones.

Rhizo'physis (ριζα a root, φυω to produce). An appendage at the extremity of certain roots.

Rhizopo'dium (ριζα a root, πους a foot). The Mycelium of Fungi.

Rhizosper'meæ (ριζα a root, σπερμα a seed). Synonyme for Marsiliaceæ.

Rhi'zula (diminutive formed from ριζα a root). The roots emitted by the sporules of Musci, and some other Acotyledones.

Rhododen'dra, Rhododendre'æ, Rhodora'ceæ (from the genus Rhododendron or Rhodora). Synonymes for Ericaceæ.

Rhodo-leu'cus (ροδον a rose, λευκος white). Combination of red and white.

Rhombifo'lius (*rhombus* a rhomb, *folium* a leaf). Where the leaf is rhomboidal.

Rhom'boid, Rhomboi'dal, *Rhombe'us*, *Rhombifor'mis*, *Rhomboi'deus*, *Rhomboida'lis* (*rhombus* a rhomb, *forma* shape). Rudely approximating to the form of a rhomboid; that is to say, to a quadrangular figure (not a square), whose sides are equal. Fig. 138.

138

Rhynchos'porus (ρυγχος a beak, σπορα a seed). Where a fruit terminates in an elongated projection.

Rib. Any marked nerve in the leaf; but more especially the central longitudinal one.

Rib'bed. Where strongly-marked nerves (one or more) proceed from the base, or near the base, to the apex of the leaf.

Ribe'sieæ (from the genus Ribes). Synonyme for Grossulaceæ.

Ric'tus (a grinning). Synonyme for *faux* in personate flowers.

Ridge. An elevated line on the cremocarp of Umbelliferæ, of which, in some cases, there are ten, termed primary, and eight secondary.

Right-handed. See Dextrorsum.

Rig'id, *Ri'gidus* (stiff). With slight or no flexibility.

Ri'ma (a cleft). An ostiolum which has the appearance of a small cleft or chink.

Rima'tus, *Rimo'sus* (full of clefts), *Rimulo'sus* (*rimula* a little chink). Where a surface is covered with cracks or fissures.

Ring. An elastic band upon the thecæ of many ferns. The line where the operculum separates from the peristome in the thecæ of mosses. The debris left round the stipes of some agarics by the bursting of the volva.

Ring'ed. When a cylindrical part is surrounded by lines, bands, elevations, &c., which approximate to circles.

Rin'gent, *Rin'gens* (grinning). A bilabiate corolla whose lips are widely separate. This includes both personate and labiate forms. Applied, also, to those florets of Compositæ whose lips contain four-fifths and one-fifth of the corolla respectively.

Ringentiflo'rus (*ringens* grinning, *flos* a flower). The receptacle of such Compositæ as bear ringent florets.

78.

Ringentifor'mis (*ringens* grinning, *forma* shape). When the florets of Compositæ approach the ringent condition.

Ripa'rius (belonging to banks). A plant whose natural station is the sides or banks of water courses.

Rivula'ris (*rivulus* a little brook). Either growing in water courses, or on their banks.

Root. A descending axis, the development of the radicle, with or without subordinate branches or fibres; most frequently subterranean, serving both to attach plants to their support, and to imbibe nourshment.

Rooting. Emitting roots.

Root-stock. A subterranean or prostrate stem, which emits roots from its lower surface.

Rope-shaped. Synonyme for Funiliform.

Ro'ridus (*ros, roris* dew). Dewy.

Ro'saceæ (from the genus Rosa). The Rose tribe. A natural order of Dicotyledones.

Rosa'ceous, *Rosa'ceus* (belonging to Roses). When parts, more or less laminated, are arranged in a whorl round an axis, in a manner somewhat resembling the disposition of the petals of a rose. Belonging to the natural order Rosaceæ.

Rosela'tus. Synonyme for Rosaceus.

Rosel'la (diminutive of Rosa, a rose). A small terminal aggregation of leaves or other appendages, with a rosaceous arrangement.

Rose'us (rose-colour). Pale red.

Rostella'tus (*rostellum* a little beak). Furnished with a stiff and often somewhat hooked termination.

Rostel'lum (a little beak). An extension of the upper edge of the stigma in some of the Orchideæ. A hook.

Rostra'tus. Beaked. Synonyme for Rostellatus.

Ro'strum (a beak). Synonyme for Rostellum. The term is extended also to other beak-like prominences.

Ro'sula. Synonyme for Rosella.

Rosula'ris, Rosula'rius. Having the arrangement described under Rosella.

Rosula'tus. Synonyme for Rosaceus.

Ro'tate, *Rota'ceus, Rotæfor'mis, Rotif'ormis, Rota'tus* (*rota* a wheel, *forma* shape). When a monopetalous corolla has a very short tube, and spreading limb. Fig. 139.

139

Rota'tion, *Rota'tio* (a wheeling about). The internal circulation of the fluids in the cells of plants.

Rotunda'tus (made round). Roundish.

Rotun'dus. Round. Synonyme for *Orbicularis.* Sometimes used synonymously with Rotundatus.

Rough. Where the surface is covered with hard but small elevations; and also where it is coated with stiff hairs.

Roughish. Approaching the character of Rough, but not being decidedly so.

Round. Synonyme for Orbicular.

Roundish. Approaching a round form, or merely rounded off at the extremities.

Roxburghia'ceæ (from the genus Roxburghia). A natural order of Monocotyledones.

Rubel'lus (somewhat red). *Rubes'cens* (growing red), *Ru'bens* (red), *Rubicun'dus* (ruddy). Various modifications of *Ruber.*

Ru'ber. Pure red of a deep tint.

Rubico'lus (*Rubus* a genus of Rosaceæ, and *colo* to inhabit). Parasitic on, or attached to, the stems or leaves of a Rubus.

Rubigin'eus, Rubigino'sus (rusty). Of a brownish red tint. Red with much grey.

Rudera'lis (*rudis* rubbish). Growing in waste places, or among rubbish.

Ru'dimentary. Either in an early stage of development, or in an imperfectly developed condition.

Ru'fescens, Ru'fus (somewhat red). Brown inclining to red. Red with still more grey than in Rubigineus.

Ru'ga. A wrinkle.

Rugo'se, *Rugo'sulus, Rugo'sus* (rough, wrinkled), *Rugulo'sus.* Where a surface is covered with wrinkles.

Rumina'ted, *Rumina'tus* (chewed), When the hard albumen of some seeds (as the Nutmeg, fig. 140) is penetrated by irregular channels, filled with softer cellular matter.

Run'cinate, *Runcina'tus,* (*runcina* a large saw). Where the large marginal incisions of a leaf are directed in a curved and serrated manner towards the base. Fig, 141.

Run'ner. A slender prostrate stem, rooting at the joints.

Rupes'tris, Rupic'olus (*rupes* a rock, *colo* to inhabit). Growing naturally on rocks.

Rup'tilis (*ruptus* broken). Bursting irregularly, without any defined line of dehiscence.

Ruptiner'vis, *Ruptiner'vius* (*ruptus* broken, *nerva* a nerve). Where uniformity in the size of a leaf-nerve is interrupted by swellings.

Rup'turing. Irregularly bursting.

Rura'lis (rural). Growing in situations peculiar to country places; as on the thatch of a cottage.

Rusty. Synonyme for Ferruginous.

Ru'taceæ, Ru'tæ, Ru'teæ (from the genus Ruta). The Rue tribe. A natural order of Dicotyledones.

Ru'tilans Ru'tilus (fiery, red). Of a brick-red colour. Red with a moderate portion of grey. Also glittering with red.

Ry'tidocarpus (ρυτις a wrinkle, καρπος fruit). Where the surface of the fruit is covered with wrinkles.

Sabuli'colus (*sabulum* sand, *colo* to inhabit), *Sabulosus* (sandy). Growing in sandy places.

Sac. A vesicle in the nucleus, within which the embryo is formed.

Sacchara'tus, *Sacchari'nus* (*saccharum* sugar). Having a sweet taste.

Sac'ciform, Saccifor'mis (*saccus* a sack, *forma* shape). Having the general form of a sack.

Sac'culus (a little sack). The peridium of some Fungi.

Sac'cus (a sack). The sac. Synonyme for Corona.

Sacel'lus, (*sacellum* a chapel). A one-seeded indehiscent pericarp, invested by the hardened perianth.

Sack. See Sac.

Saddle-shaped. Bending down at the sides, so that a rounded form is given to the upper part.

Saffron-coloured. Deep orange, with a very slight admixture of grey.

Sa'gittate, *Sagitta'lis*, *Sagitta'tus* (shot with an arrow). Pointed at the apex, and the base prolonged backwards from the sides into two acute ears. Fig. 142.

Salica'ceæ, Salici'neæ (from the genus Salix). The Willow tribe. A natural order of Dicotyledones.

Sali'nus (*salina* a salt pit), *Sal'sus* (salted). Tasting of salt. Synonymes for Salsuginosus.

Salsugino'sus (*salsugo* a salt liquor). Growing in salt places, like marshes by the sea.

SALVADORA'CEÆ (rom the genus Salvadora). A natural order of Dicotyledones.

SALVER-SHAPED. Synonyme for Hypocrateriform.

SALVINIA'CEÆ, SALVINI'EÆ (from the genus Salvinia). A natural order of Acotyledones.

SA'MARA (rather *SAMERA*, an elm seed). A compressed, few-seeded, coriaceous or membranaceous indehiscent pericarp, with a membranaceous expansion at the end or edges. Fig. 143 is one half the fruit of the Sycamore.

SAMA'ROID, ***SAMAROI'DEUS*** (from *SAMARA*, and ειδος form). Resembling a Samara.

SAMYDA'CEÆ, SAMYD'EÆ (from the genus Samyda). A natural order of Dicotyledones.

SAN'GUINE, ***SANGUI'NEUS*** (blood-colour). Red with much grey.

SANGUISORBA'CEÆ, SANGUISOR'BEÆ (from the genus Sanguisorba). The Burnet tribe. A natural order of Dicotyledones. Otherwise considered as a sub-order of Rosaceæ.

SANTALA'CEÆ (from the genus Santalum). The Sanders-wood tribe. A natural order of Dicotyledones.

SAP, ***SA'PA*** (sodden wine). A general term for the juices of a plant. The ascending sap is the crude material introduced by absorption; the descending sap (called proper juice) is the elaborated material, which then contains organized compounds, suited to the nutrition of the plant.

SA'PIDUS (savory, from *SAPOR* a taste). Possessing a pleasant taste.

SAPINDA'CEÆ. SAPIN'DI (from the genus Sapindus). The Soap-tree tribe. A natural order of Dicotyledones.

SAPONA'RIUS (*SAPO* soap). Possessing detergent properties, like soap.

SAPOTA'CEÆ, SAPO'TÆ, SAPO'TEÆ (from Achras Sapota). The Sappodilla tribe. A natural order of Dicotyledones.

SARCOBA'SIS (σαρξ flesh, βασις a base). Synonyme for Gynobasis, when very fleshy. Synonyme for Carcerulus.

SAR'COCARP, ***SARCOCAR'PIUM*** (σαρξ flesh, καρπος fruit). The intermediate and more succulent part of the pericarp which lies between the epicarp and the endocarp.

SARCODER'MA, SARCODER'MIS (σαρξ flesh, δερμα skin). A layer more or less apparent, and sometimes fleshy, between the Exopleura and Endopleura.

Sarcoi'des (σαρξ flesh, ειδος resemblance). Having the general appearance of a piece of flesh.

Sarcolo'beæ (σαρξ flesh, λοβος a lobe). The most extensive of two primary groups into which the Leguminosæ may be divided, characterized by the cotyledones being thick and fleshy.

Sarco'ma (σαρξ flesh). A fleshy disk.

Sarmenta'ceous, *Sarmenta'ceus* (rather *Sarmentitius* belonging to twigs). When a branch approaches to, or assumes the character of a runner.

Sarmentif'erus, *Sarmento'sus* (full of twigs). Bearing long flexible branches, which require support to prevent their trailing on the ground.

Sarmen'tum (a twig). A runner.

Sarracenia'ceæ, Sarrace'nieæ (from the genus Sarracenia). A natural order of Dicotyledones,

Saturate-virens (*saturatus* full of a deep colour). Of a grass-green; where the green tint appears full, without admixture.

Saurura'ceæ, Sauru'eæ (from the genus Saururus). A natural order of Dicotyledones.

Sausage-sha'ped. A cylindrical tube with nearly hemispherical terminations; fig. 144.

Santel'lus (*santelles* an attendant). Synonyme for Bulbillus.

Sawed. Synonyme for Serrate.

Saxa'tilis (living among rocks), *Saxi'colus* (*saxum* a rock, *colo* to inhabit), *Saxo'sus* (stony). Growing spontaneously in rocky and stony stations.

Saxifraga'ceæ, Saxi'fragæ, Saxifra'geæ (from the genus Saxifrage). The Saxifrage tribe. A natural order of Dicotyledones.

Sca'bridus, *Scabrius'culus* (*scaber*, rough). Somewhat rough or harsh to the touch.

Scabri'ties (scabbiness). Minute scaly pubescence producing a roughness of the surface.

Sca'brous, *Sca'ber* (rough). Where a sensation of harshness or roughness is produced by stiff pubescence, or scattered tubercles.

Scævola'ceæ, Scævo'leæ, (from the genus Scævola). A natural order of Dicotyledones.

Scala'riform, *Scala'riformis* (*scalaris* a ladder, *forma* shape). When vasculai tissue is transversely striated, as if barred, like the steps to a ladder

SCALE. The degenerate or rudimentary state of leaves resembling a fish-scale; more especially such as those that form the outer portions of a bud. Very generally and vaguely applied to a variety of small membranous expansions. *LEPIS* has been sometimes restricted to small peltate scales; and *SQUAMA* reserved for those attached by one extremity to their support.

SCA'LY. Furnished with scales.

SCALPEL'LIFORMIS (*SCALPELLUM* a little knife, or lancet, *FORMA* shape). Shaped like the blade of a penknife, and often (as among phyllodia) with the surfaces set vertically with respect to the axis of vegetation; fig. 145.

SCAN'DENS (climbing). When stems which would otherwise trail upon the ground are raised by the support of tendrils, claws, &c.

SCAPE, *SCA'PUS* (a stem). A peduncle, rising from a depressed or subterranean stem, with the lower internode very long, and consequently with few or no bracts, except near the summit, where one or more pedicels originate. A synonyme for the stipe of some Fungi.

SCAPEL'LUS (diminutive of Scapus). The neck or caulicule of the germinating embryo.

SCAPHI'DIUM (*SCAPHIUM* a hollow vessel). The spore-case of Algæ.

SCA'PHIUM (a hollow boat-like vessel). Synonyme for the Carina of a papilionaceous flower.

SCAPIFLO'RUS (*SCAPUS* a scape, and *FLOS* a flower). Having the flowers on scapes.

SCAPIFOR'MIS, *SCAPIG'ERUS* (*SCAPUS* a scape, *FORMA* shape, *GERO* o bear). Where a stem, being defective in leaves, assum-- the appearance of a Scape.

SCAR. See Cicatrix.

SCARIO'SE, SCARI'OUS, *SCARIO'SUS* (*SCARROSUS?* rugged). Thin, dry, and membranous.

SCARRED. Marked by scars.

SCARROSE. Synonyme for Squarrose.

SCATTERED. Without apparent symmetry in arrangement.

SCEPA'CEÆ (from the genus Scepa). A natural order of Dicotyledones.

Schista'ceus, *Schisto'sus* (*schistos* slate-stone). Of the colour of common roofing slate. Blue with much grey.

Scimitar-shaped. Synonyme for Acinaciform.

Sci'on. The young state of a branch whilst closely invested with leaves in the form of scales.

Scitami'neæ (*scitamentum* choice food). The Ginger tribe. A natural order of Monocotyledones.

Sciuroi'des (σκίουρος a squirrel, ειδος resemblance). Curved and bushy like a squirrel's tail.

Scleranthа'ceæ, Scleran'theæ (from the genus Scleranthus). A natural order of Dicotyledones.

Scleran'thum (σκληρος hard, ανθος a flower). Synonyme for Diclesium.

Scler'oid, *Scleroi'dus* (σκληρος hard). Of a hard texture.

Scle'rogen (σκληρος hard, γενεσίς creation). A non-nitrogenized compound which fills the cells of woody fibre; and forms the hardened bony matter in some fruits.

Sclerophyl'lus (σκληρος hard, φυλλον a leaf). With the leaves stiff and hard.

Scleropoi'dus (σκληρος hard, πους foot). When persistent peduncles harden and become thorny.

Scobicula'tus, *Scobifor'mis* (*scobs* sawdust, *forma* shape). In fine grains like sawdust.

Scobi'na (a file).. The immediate support to the spikelets of grasses.

Scobina'tus, (*scobina* a file). Where the surface feels rough like a rasp.

Scorpio'id, Scorpioi'dal, *Scorpioi'des*, *Scorpioida'lis* (σκορπιος scorpion, ειδος resemblance). Where a main axis of inflorescence is curved in a circinate manner, like the tail of a scorpion; fig. 146.

146

Scrobicula'tus *Scrobiculo'sus* (*scrobiculus* a little ditch). Pitted.

Scrophulari'aceæ, Scrophula'riæ, Scrophulari'neæ, (from the genus Scrophularia). The Fig-wort tribe. A natural order of Dicotyledones.

Scro'tiform, *Scrotifor'mis* (*scrotum* a bag, *forma* shape). Pouch-shaped.

Scurf. Minute scales of membranous matter on the surface of some tissues. See Scale.

SCURFINESS. The appearance produced by membranous superficial scales.

SCU'TATE, *SCUTA'TUS* (armed with a shield). Synonyme for Buckler-shaped.

SCUTEL'LA (a dish or saucer). Also, *SCUTELLUM*. A sessile Apothecium bordered by the substance of the thallus itself.

SCUTELLA'RIS, *SCUTELLA'TUS* (*SCUTELLA* a saucer). When branched hairs are combined into saucer-shaped disks. Also when a thallus is covered with scutellæ.

SCUTEL'LIFORM, *SCUTELLIFOR'MIS* (*SCUTELLA* a platter, *FORMA* shape). Somewhat oval, disk-like, and concavo-convex; fig. 147.

147

SCUTEL'LUM (diminutive for *SCUTUM* a shield). An apothecium with an elevated rim formed by the thallus.

SCU'TIFORM, *SCUTIFOR'MIS* (*SCUTUM* a shield, *FORMA* shape). Synonyme for Buckler-shaped.

SCU'TUM (a shield). A circular disk-like space over the stigma, in the midst of the orbiculus, in some plants.

SCY'PHA, *SCY'PHUS* (a large cup). A cup-shaped podetium.

SCYPHIFOR'MIS (*SCYPHUS* a large cup, *FORMA* shape). Cup-shaped.

SCY'PHULUS (diminutive for *SCYPHUS* a large cup). The cup-like appendage from which the seta of Hepaticæ arises. Used also synonymously with Scypha.

SCY'PHUS (a large cup). A funnel-shaped corona. Used also synonymously with Scypha, Scyphulus, and Pyxidium.

SCYTI'NUM (σκυτινος made of leather). A tough form of legume, woody externally, and pulpy within.

SEA-GREEN. Synonyme for Glaucous.

SEBA'CEUS (for *SEVACEUS*, from *SEVUM* tallow). Looking like lumps of tallow.

SEBIF'ERUS (*SEBUM*, for *SEVUM* tallow). Producing vegetable wax.

SECRE'TION, *SECRE'TIO* (a separating). A vital function by which abstraction is made of some portion of the constituents of a nutritive fluid. The residium left after such abstraction is also termed a secretion.

SEC'TILE, *SEC'TILIS* (easy to be cut). Subdivided into small portions.

SECT'US (cut). Parted.

SECU'ND, *SECUNDA'TUS*, *SECUN'DUS* (next in the same rank). When flowers, or particular organs, are all turned to the same side of the axis round which they are arranged; fig. 148.

SECUNDIFLO'RUS (*SECUNDUS* secund, and *FLOS* a flower). Where the flowers are secund.

SECUN'DINE, *SECUNDI'NA* (*SECUNDUS* second). The inner or first-developed integument to the nucleus of the ovule. Synonyme for Regmen.

SEED. The fertilized ovule.

SEGETA'LIS (*SEGES* a corn field). Growing among corn.

SEG'MENT, *SEGMEN'TUM* (a parting). One of the subdivisions of any part or organ.

SEGREGA'TE (*SEGREGATUS* separated). An Order of the artificial Linnean class Syngenesia, in which several one or few-flowered capitula are closely aggregated into a compound capitulum.

SELAGINA'CEÆ, SELAGI'NEÆ (from the genus Selago). A natural order of Dicotyledones.

SELLÆFOR'MIS (*SELLA* a saddle, *FORMA* shape). Saddle-shaped.

SE'JUGUS (*SEX* six, *JUGUM* a yoke). In six pairs; as in some pinnate leaves.

SE'MEN. Seed.

SE'MI (half). In composition, (with other terms,) generally implies a partial or imperfect exhibition of the particular effect implied by the term with which it is compounded. As in several of the following instances.

SEMI-ADHE'RENS. The adhesion extending through a portion of the usual condition from below upwards.

SEMI-AMPLEC'TUS, *SEMI-AMPLECTI'VUS*. May be used synonymously with Equitans.

SEMI-AMPLEXICAU'LIS, *SEMI-AMPLEX'US*. Half-clasping.

SEMI-CAP'SULA. Synonyme for Cupula.

SEMI-COLUMNA'RIS (*COLUMNA* a pillar). Synonyme for Semiteres.

SEMI-CORDA'TUS, *SEMI-CORDIF'ORMIS*. Cordate on one side only of the longitudinal axis.

SEMI-CYLINDRA'CEUS, *SEMI-CYLIN'DRICUS* (*CYLINDRUS* a cylinder). Synonymes for Semi-teres.

SEMI-DI'GYNUS. When two carpels cohere near the base only.

SEMI-DOUBLE, *SEMI-DU'PLEX*. When the innermost stamens continue perfect, whilst the outermost have become petaloid.

SEMI-FLOS'CULAR, *SEMI-FLOSCULO'SUS*, *SEMI-FLOS'CULUS*. When all the florets (in Compositæ) are ligulate.

SEMI-LOCULA'RIS, *SEMI-LO'CULUS* (*SEMI* half, *LOCULARIS* with loculaments). Where the dissepiments are incomplete, and consequently the pericarp is really unilocular.

SEMI-LUNA'TUS (*SEMI* half, *LUNATUS* like a half moon). Synonyme for Lunatus.

SE'MINAL, *SEMINA'LIS* (bearing reference to sowing, or to the seed). Whatever has reference to some portion of the seed.

SEMINA'TIO (the act of sowing). Dissemination.

SEMINIF'ERUS (*SEMEN* a seed, *FERO* to bear). Bearing reference to the particular portion of the pericarp to which the seeds are attached. Has been used synonymously with dicotyledonous.

SEMINIFOR'MIS (*SEMEN* a seed, *FORMA* shape). Applied to certain reproductive bodies among Acotyledones, which are not parts of fructification.

SEMIN'ULA, *SEMIN'ULUM* (diminutive for *SEMEN* a seed). Synonyme for Spora.

SEMINULIF'ERUS (*SEMINULA* and *FERO* to bear). The portion of a cryptogamic plant which bears the spores. Applied also to the cavity of the ovarium in its early state, when the ovules are yet unfertilized.

SEMI-ORBICULA'TUS (*SEMI* half, *ORBICULATUS* rounded). Hemispherical.

SEMI-OVA'LIS (*SEMI* half, and *OVALIS*). Where the portion on one side of the longitudinal axis is oval, but not that on the other side.

SEMI-OVA TUS (*SEMI* half, and *OVATUS*). Where the portion on one side of the longitudinal axis is ovate, but not that on the other side.

SEMI-PETALOI'DEUS (*SEMI* half, and *PETALOIDEUS*). Synonyme for Petaloideus.

SEMIRAD'ICANS, *SEMIRADIA'TUS* (*SEMI* half, and *RADIATUS*). When only a portion of the outer florets of a capitulum have the corolla radiant, or differently formed from those of the disk.

SEMI-RENIFOR'MIS (*SEMI* half, and *RENIFORMIS*). Where the portions lying on one side only of the longitudinal axis is reniform.

SEMI-RETICULA'TUS (*SEMI* half, *RETICULATUS*). When the

outer one of several layers is reticulate, the rest membranous.

SEMI-SAGITTA'TUS (*SEMI* half, and *SAGITTATUS*). Where the portions lying on one side only of the longitudinal axis is sagittate.

SEMI-SEPTA'TUS (*SEMI* half, and *SEPTATUS*). Where projections into a cavity do not extend sufficiently far to subdivide it into separate cells; fig. 149.

149

SEMI-STAMINA'RIS, *SEMI-STAMINA'RIUS* (*SEMI* half, *STAMINARIS*). Semi-double, by the transformation of a portion of the stamens into petals.

SEMI-SYMPHIOSTEM'ONIS (*SEMI* half, συμφυω to unite, στημων a stamen). Where a portion of the stamens cohere, the rest remaining free.

SEMI-TE'RES (*SEMI* half, and *TERES*). One side cylindrical the other flat. Half-terete.

SEMI-VALVA'TUS, *SEMI-VAL'VIS* (*SEMI* half, and *VALVATUS*). Where the valves are only partially dehiscent.

SEMPERVI'RENS (*SEMPER* always, *VIRENS* green). With green leaves or surface throughout the year.

SENA'RIUS (containing six), *SE'NI* (by sixes). Arranged in six together of the same kind.

SENSI'BILIS (*SENTIO* discernable by the senses). Sensitive.

SENSITIVE. Manifesting "irritability."

SE'PAL, *SE'PALUM* (by substituting σ for π in πεταλον). One of the foliaceous expansions forming the subordinate parts of the calyx.

SEPALI'NE, SE'PALOUS, *SEPALI'NUS* (from *SEPALUM*). Having reference to sepals.

SEPALOI'D (*SEPALUM* a sepal, εἶδος resemblance). Looking like a sepal.

SEPA'LULUM (diminutive for *SEPALUM*), Sometimes applied to a subordinate part of a calyculus, or the accessary bracts about a calyx.

SEP'ARATE. Without cohesion or adhesion to neighbouring parts.

SEPI'COLUS (*SEPES* a hedge, *COLO* to inhabit). Whose natural habitat is in hedgerows and copses,

SEPTA'LIS (see *SEPTUM*). Belonging to a Septum.

SEPTA'TUS (see *SEPTUM*), Possessing Septa.

SEPTENA'TUS, *SEPTE'NUS* (*SEPTENI* seven), Where either subordinate parts, or the subdivisions of one part, amount to seven,

SEPTICI'DAL, *SEPTICI'DUS* (*SEPTUM* and *CÆDO* to cut). Where dehiscence takes place along the lines of suture (*a*), or separates the dissepiments (*b*) formed by contiguous carpels of compound fruits; fig. 150.

SEPTIF'ERUS (*SEPTUM* and *FERO* to bear). When some part supports certain portions of tissue which serve to subdivide (partially or wholly) some other part.

SEPTIFOR'MIS (*SEPTUM* and *FORMA* shape). Having the general form and appearance of a dissepiment, though not strictly such.

SEPTIF'RAGAL, *SEPTIF'RAGUS* (*SEPTUM* and *FRAGO* to break). Where dehiscence takes place along the lines of suture, and at the same time the valves separate from the dissepiments, which are not subdivided as in the septicidal dehiscence; fig. 151.

SEP'TILIS (*SEPTUS* enclosed). Having relation to the Septum.

SEPTULA'TUS (*SEPTUS* enclosed). Furnished with spurious transverse dissepiments; fig. 152, *d*.

SEP'TULUM (diminutive of *SEPTUM*). When a partition is small.

SEP'TUM (a hedge). A partition of any kind; but more especially one which subdivides the ovary or fruit, and originates in the union of a portion of the contiguous carpels extending to the axis; *s*, in fig. 153.

SEPTUPLINER'VIS, *SEPTUPLINER'VIUS* (*SEPTEMPLEX* sevenfold, *NERVUS* a nerve). Where three strong nerves are given off on either side of the midrib. See Quinquenerved.

SERIA'LIS (*SERIES* a row). Arranged in rows.

SERI'CEUS (*SERICUS* made of silk). Covered with fine, rather long, close and soft hair, which has a silky appearance.

SERO'TINUS (happening late). When a plant flowers later in the year than others to which it is related.

SERRÆFO'LIUS, *SERRATIFO'LIUS* (*SERRA* a saw, *FOLIUM* a leaf). Having serrated leaves.

SER'RATURE, *SERRATU'RA* (*SERRA* a saw). Synonyme for tooth, when applied to the incisions on the margins of leaves, &c.

SER'RATED, *SERRA'TUS* (sawed). When marginal serratures are sharp and pointed forward, looking like the edge of a saw; fig. 154.

SERRULA'TUS (*SERRULA* a little saw). Synonyme for Denticulatus.

Ser'tulum (diminutive for *sertum* a garland). Synonyme for Umbella, when simple.

Sesqui-alter (containing one and a half), Where there is half as much more as the number of some other part to which a given part bears special relation; as where the stamens are one half as many more as the petals or sepals. Where a fertile flower is accompanied by an abortive one, as in some Grasses.

Sessi'le, *Ses'silis* (as it were sitting). Where an organ is attached to its support without the intervention of some intermediate part; as when a leaf is without petiole (*sessilifolius*); a flower without a pedicel (*sessiliflorus*).

Se'ta (a bristle). Any stiff hair. The stalk supporting the theca of Mosses. The arista of Grasses, when it is not below the apex but forms a termination to any of the floral bracts.

Seta'ceo-serra'tus (*setaceus* and *serratus*). Where the serratures taper off to bristle-like points.

Seta'ceous, *Seta'ceus* (*seta* a bristle). Possessing the characters ascribed to a seta.

Setifor'mis (*seta* a bristle, *forma* shape). Shaped like a seta.

Setig'erus (*setiger* bearing bristles). Furnished with one or more setæ.

Seto'se, *Seto'sus* (full of bristles). Covered with setæ. Used also synonymously with Setigerus.

Se'tula (diminutive for *seta* a bristle). The stipes of certain Fungi.

Sex, *Sex'us* (a sex). When stamens or pistils alone, or when both these organs are formed in flowering plants, the flowers are termed male, female, or hermaphrodite accordingly.

Sexan'gular, *Sexangula'ris* (*sexangulus* six-cornered). With six angles, or decided projections from the surface.

Sexfa'rius (*sexfariam* six manner of ways). Presenting six rows, extending longitudinally round an axis.

Sextu'plex (*sex* six). Where a part is six times repeated.

Sex'ual, *Sexua'lis* (*sexus* a sex). Having some reference to a distinction of sex.

Shaggy. Where the pubescence is composed of long but not stiff hairs.

Sharp-pointed. Synonyme for Acute.

Sheath, A petiole, or a portion of it, which embraces the stem to which it is attached; fig. 155.

155

Shield. A cup-like expansion or disk on the thallus of Lichens, which contains asci.

Shield-shaped. Synonyme for Scutate.

Shining. When a surface is smooth and polished.

Shoot. Any fresh branch, more especially one given off immediately from the upper extremity of the root.

Short. Not so long as some neighbouring part to which reference is intended to be made.

Shrub. Woody plants which do not form a true trunk like trees, but have several stems rising from the roots.

Sic'cus (dry). Containing little or no aqueous matter.

Sigilla'tus (impressed by a seal). When a rhizoma is marked by scars left by the fall of branches successively developed upon it.

Sigmoi'd, ***Sigmoi'deus*** (σιγμα the letter S, ειδος resemblance). Curved in two directions like the letter S.

Silena'ceæ, **Sile'neæ** (from the genus Silene). A natural Order, or else a subordinate Group of Caryophyllaceæ.

Si'licle, ***Silic'ula*** (a little pod). Formed like a Siliqua, but where the length never exceeds four times the breadth.

Siliculo'sa. A Linnean order of the class Tetradynamia, containing plants whose seed vessels are siliculæ.

Siliculo'sus (*silicula* a silicle). Possessed of a silicle; or resembling one.

Sil'iqua (a Bean pod). A dry bivalvular fruit, generally with a transverse, membranous and spurious dissepiment, formed by the extension and union of the opposite placentæ.

Siliquel'la (diminutive from *Siliqua*). A subordinate part of such fruit as the Poppy, composed of the carpel and two extended placentæ; fig. 156 is a section.

156

Siliquo'sa. Linnean order of the class Tetradynamia, containing plants whose seed-vessels are siliquæ.

Siliquo'sus (*siliqua* a Bean pod). Where the fruit is a siliqua, or resembles one.

Silky. When hairs are very long and fine, with a glossy appearance like silk.

Silver-grain. Popular name for medullary rays.

Silvery. White, slightly tinged with bluish grey, and possessing a metallic lustre.

Simaruba'ceæ (from the genus Simaruba). The Quassia tribe. A natural order of Dicotyledones.

SIMILIFLO'RUS (*SIMILIS* like, *FLOS* a flower). When an umbel has all its flowers alike.

SIMPLE, *SIM'PLEX*. In opposition to "compound," where there are no subordinate parts or distinct ramifications.

SIMPLICIS'SIMUS (very simple). Without the slightest tendency to subdivisions or ramifications.

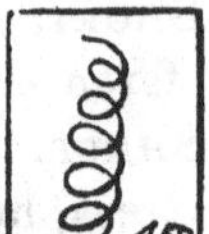

SINIS'TRORSE, *SINISTROR'SUM* (towards the left hand). Where the coils of a spiral would appear, to a person in the axis, to rise from right to left; fig. 157.

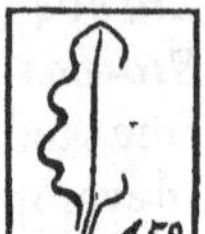

157

SIN'UATE, SINUA'TED, *SINUA'TUS* (crooked). A margin rendered uneven by alternate rounded, and rather large, lobes and sinuses; fig. 158.

158

SINUATO-DENTA'TUS. Between sinuate and dentate.

SINUOLA'TUS (diminutive of *SINUATUS*). Synonyme for Repandus.

SI'NUS, *SI'NUS* (a bay). The re-entering angle or depression between two projections or prominences.

SIPHONI'PHYTUM (σιφωνιον a siphon, φυτον a plant). A composite plant with all the florets floscular.

SI'TUS (situation). The peculiar mode in which parts are disposed, as well as the position they occupy.

SLASHED. Where a surface is divided by deep and very acute incisions; fig. 159.

159

SLATE-GREY. Bluish grey. Blue with a large admixture of grey.

SLEEP. A peculiar vital effect produced on some expanded flowers, and the leaflets of certain leaves; by which they become closed or folded together at certain periods of the day.

SLENDER. Long and thin.

SLI'MY. See Mucous.

SMALL. Has respect to something with which comparison in size is supposed to be made.

SMARAG'DINUS (like an emerald). Pure green without any admixture of grey.

SMILA'CEÆ (from the genus Smilax). The Smilax tribe. A natural order of Monocotyledones.

SMO'KY. Dull and very dark grey.

SMOOTH. Devoid of any kind of uneveness.

SNOW-WHITE. Perfectly pure white.

SO'BOLES. A shoot.

SOBOLIF'ERUS (*SOBOLES* a shoot, *FERO* to bear). Bearing shoots,

So'cial, *Socia'lis* (belonging to allies or confederates). When many individuals of the same species usually grow together in a wild state, so as to occupy a considerable extent of ground.

Soft. When a part is composed of tissue which yields readily to the touch.

Solana'ceæ, Sola'neæ (from the genus Solanum). The Night-shade tribe. A natural order of Dicotyledones.

Sol'idus (solid). Without cavities of any kind.

Sol'itary, *Solita'rius* (alone). Not closely associated with another object of the same description.

Solubil'ity, *Solubil'itas*. The property of separating into distinct portions by a kind of spurious articulation; as when certain legumes become transversely divided between the spaces occupied by the seeds; fig. 160.

160

Solu'tus (loosed). Separate.

Som'nus. Sleep.

Sooty. See Fuliginosus.

Sor'didus (dirty). When a colour contains more or less admixture of grey. *Sordidis'simus*, when the grey greatly predominates.

Sore'dium, *Sore'uma* (σωρὸς a heap). A patch of Propagula (otherwise termed Gonidia) which have burst through the surface of the thallus of Lichens.

Soro'sa, *Soro'sis*, *Soro'sus* (σωρευσις a heaping up). A compound fleshy fruit formed by the close aggregation of many flowers whose floral whorls become succulent.

So'rus (σωρος a heap). A patch of the aggregated thecæ in Ferns.

Spadi'ceus (*spadix* some red colour). Bay. Clear reddish-brown. Red with a small admixture of grey.

Spadici'neæ (from *spadix*). A group proposed to contain such Orders as have their flowers arranged on a spadix.

Spa'dix (a Phœnician musical instrument). The axis of a spiked inflorescence among monocotyledones, when the flowers are densely aggregated. It is usually, but not always, accompanied by one or more spathes, and is frequently fleshy.

Span. See Dodrans.

Spanan'thus (σπανος rare, ανθος a flower). Bearing few flowers.

SPAR'SUS (scattered). Irregularly, and often scantily, distributed; as *SPARSIFLORUS*, having a few scattered flowers; *SPARSIFOLIUS*, where the leaves are distantly scattered over the herbage.

SPA'THA. A spathe.

SPATHA'CEÆ (from spatha). A Linnean group of certain Monocotyledones, furnished with a spathe.

SPATHA'CEUS (from *SPATHA* a spathe). Either furnished with a spathe, and more especially if it is large; or, having the general appearance of a spathe.

SPATHE (*SPATHA* the flowering branch of the Date; *SPATHE* a tree resembling a Palm). A foliaceous or membranaceous involucrum, of one or few sheathing bracts, in certain Monocotyledones, which more or less envelope the flowers.

SPATH'ELLA (diminutive of *SPATHA*). Synonyme for Gluma; also extended to Palea in Grasses.

SPATHEL'LULA (diminutive for *SPATHELLA*). Synonyme for Palea, in Grasses.

SPA'THULATE, SPA'TULATE, *SPATHULA'TUS* (*SPATHULA* a spoon). More or less rounded towards the summit, and narrowed towards the base; fig. 161.

161

SPE'CIES, *SPE'CIES* (a form). An assemblage of forms which (it is empyrically assumed) might have emanated, according to the laws of reproduction, from one or more individuals of a particular form, as this was impressed by the Creator, when such form was first called into existence.

SPEIRE'MA (σπειρημα a seed). Synonyme for "Propagulum" in Lichens; otherwise called "Gonidium."

SPERMAN'GIUM (σπερμα seed). The spore-case of Algæ.

SPERMAPH'ORUM, *SPERMOPH'ORUM* (σπερμα seed, φερω to bear). Synonyme for Placenta; and also for Funiculus.

SPERMAPO'DIUM, *SPERMAPODOPH'ORUM* (σπερμα seed, πους a foot, φερω to bear). The branched gynophorus of the Umbelliferæ.

SPERMATI'DIUM, *SPERMA'TIUM* (σπερμα seed). The spore of Algæ.

SPERMA'TO-CYSTID'IUM (σπερμα seed, κυστις a bladder). Synonyme for Anthera; and more especially for the supposed anther of Musci, otherwise called Antheridium.

SPERMATOI'DIUM (σπερμα seed, ειδος resemblance). A case containing the propagula (otherwise gonidia) in Algæ.

SPERMI'DEUS (σπερμα seed). Producing seed.

SPERMI'DIUM (σπερμα seed). Synonyme for Achenium.

SPER'MODERM, *SPERMODER'MIS* (σπερμα seed, δερμα skin). The skin or integument of a seed, formed by the union of the several coats which invested the embryo in its earlier stages.

SPERMO'PHORUM (σπερμα seed, φερω to bear). Synonyme for Placenta.

SPERMOTHE'CA (σπερμα seed, θηκη a box). Synonyme for Pericarp.

SPER'MA, *SPER'MUM* (σπερμα). The Seed.

SPHÆREN'CHYMA (σφαιρα a sphere, εγχυμος succulent; or else χεῦμα something spread out). Cellular tissue in which the separate vesicles are more or less spherical.

SPHÆROBLAS'TUS (σφάιρα a sphere, βλαστος a germ). When a monocotyledonous embryo produces a cotyledon, during germination, which terminates in a swollen globular apex.

SPHÆROCAR'PUS (σφαιρα a sphere, καρπος fruit). When a fruit is globular.

SPHÆROCE'PHALUS (σφαìρα a sphere, κεφαλη a head). Where a capitulum is globular.

SPHÆRO'PHYTUM (σφαιρα a sphere, φυτον a plant). Synonyme for Filix, a Fern; the fructification (the theca) being globular.

SPHÆROS'PORA (σφαιρα a sphere, σπορα a seed). Synonyme for Tetraspore.

SPHÆ'RULA (diminutive for *SPHÆRA* a sphere). A more or less rounded peridium, discharging at the summit its sporidia through a pore or slit.

SPHÆRULI'NUS (*SPÆRA* a sphere). Synonyme for Orbicularis.

SPHALEROCAR'PIUM, *SPHALEROCAR'PUM* (σφαλερος deceiving, καρπος fruit). A one-seeded indehiscent pericarp, invested by a persistent succulent calyx, assuming the appearance of a berry.

SPHE'RICAL, *SPHÆ'RICUS*. Closely approximating to the form of a sphere.

SPHEROI'DAL, *SPHÆROIDA'LIS* (*SPHÆROIDES*, spherical). Approximating to the form of a sphere.

SPHE'RULA. See Sphærula.

SPI'CA (an ear of corn). A spike.

SPICA'TUS, *SPICIF'ERUS*, *SPICIFLO'RUS* (*SPICA* a spike, *FERO* to bear, *FLOS* a flower). Where the flowers are disposed in a spike.

Spicifor'mis (*spica* a spike, *forma* shape). Assuming the appearance of a spike.

Spici'gerus (*spica* a spike, *gero* to bear). Synonyme for Spiciferus.

Spi'cula (diminutive from *spica*). A spikelet. Also a pointed fleshy superficial appendage (see Spiculate). Also synonyme for Acicula.

Spi'culate, *Spicula'tus* (made sharp). Where a surface is covered with fine pointed fleshy appendages. Also (from *spica* a spike) when a spike is composed of several smaller spikes (or rather spikelets) crowded together.

Spiculif'erus (*spicula* spikelet, *fero* to bear). When flowers are arranged in spikelets.

Spigelia'ceæ (from the genus Spigelia). The Wormseed tribe. A natural order of Dicotyledones.

Spike. A mode of inflorescence similar to the raceme, only the flowers have properly no pedicels; though their presence is sometimes admitted among the lowest on the axis. Where spikelets (as in Gramineæ) are arranged in close and alternating series, upon a common rachis, the inflorescence is also termed a spike; fig. 162.

162

Spikelet. A small spike, of which several, aggregated round a common axis, constitute a "compound spike". The term is more especially applied to the spiked arrangements of two or more flowers of Grasses, subtended by one or more glumes, and which are variously disposed round a common axis; fig. 163.

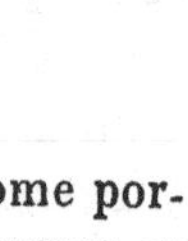

163

Spi'lus (σπιλος a stain). Synonyme for Hilum in Gramineæ.

Spi'na (a thorn). A spine.

Spindle-shaped. Synonyme for Fusiform.

Spine. A stiff sharp-pointed process, containing some portions of woody tissue, and originating in the degeneracy or modification of some organ; as of a branch, leaf, stipule, &c. It is a synonyme for thorn.

Spinel'la (diminutive for *spina* a thorn). A stout, sharp, but not ligneous, process.

Spines'cent, *Spines'cens* (*spina* a thorn). Terminating in a spine. Degenerating into a spine.

Spinif'erus (*spina* a thorn, *fero* to bear). Producing or bearing spines.

SPINIFO'LIUS (*SPINA* a thorn, *FOLIUM* a leaf). Where the leaves are spinous.

SPINIFOR'MIS (*SPINA* a thorn, *FORMA* shape). Having the general form of a spine.

SPI'NIGER (*SPINA* a thorn, *GERO* to bear). Synonyme for Spiniferus. Also for Spinescens.

SPINOCAR'PUS (*SPINA* a thorn, καρπος fruit). Where the fruit is spinous.

SPI'NOUS, *SPINO'SUS* (full of thorns). Bearing, or covered with, spines.

SPINULIF'ERUS, *SPINULO'SUS* (*SPINULA* diminutive for *SPINA*, *FERO* to bear). Furnished with very small spines.

SPINULIFLO'RUS (*SPINULA*, and *FLOS* a flower). When the sepals terminate in acute points.

SPI'RAL, *SPIRA'LIS*. Arranged in a spiral; or twisted spirally round an axis.

SPI'RAL-VESSEL. See Trachea.

SPITHAMÆ'US (*SPITHAMA* a span). About seven inches; the average space between the extremities of the thumb and forefinger when extended.

SPLEN'DENS. Glittering.

SPLIT. Divided into segments by divisions, which extend to somewhat more than half-way towards the base.

SPODO'CHROUS (σποδος ash). Of a grey tint.

SPONDIA'CEÆ (from the genus Spondias). The Hog-Plum tribe. A natural order of Dicotyledones.

SPON'GIOLE, *SPONGI'OLA* (diminutive for *SPONGIA* a sponge). The extremity of each fibre of a root, devoid of epidermis, and capable of absorbing moisture from the surrounding medium. The term is extended to the stigma (*spongiola pistillaris*), and certain parts on the surface of seeds possessing a similar property of absorption (*spongiola seminalis*).

SPON'GY, *SPONGIO'SUS*. Where the cellular tissue is copious, forming a sponge-like mass, often replete with moisture.

SPORA'DIC, *SPORA'DICUS* (σποραδικος wandering, spreading). When a given species occurs in more than one of the separate districts assigned to particular Floras.

SPORANGI'DIUM. Synonyme for Columella, in mosses. Also used synonymously with Sporangium, &c.

SPORANGI'OLUM (diminutive from *SPORANGIUM*). Synonyme for Ascus. A membranous case containing sporidia. Has been used also synonymously with Spora.

Sporangio'phorum, *Sporangiolif'erum*, *Sporan'gium*, *Sporangi'olum* (σπορα a seed, φερο and *fero* to bear). The part supporting or enclosing sporangia in certain Acotyledones.

Sporan'gium (σπορα seed, αγγος a vessel). The immediate case or covering to the spores of Acotyledones.

Spore. *Spo'ra* (σπορα a seed). A reproductive body in cryptogamous plants, analogous to the seed of Phanerogamous plants.

Spori'deus (σπορα a seed). Bearing spores. Synonyme for Acotyledoneus.

Sporidif'erus (*sporidium*, and *fero* to bear). Bearing sporidia.

Sporidifor'mis (*sporidium*, and *forma* shape). Shaped like a sporidium.

Sporidi'gerus (*sporidium* and *gero* to bear). Synonyme for Sporidiferus.

Sporidi'olum (diminutive for *sporidium*). Synonyme for Spora and Sporula, in the lower groups of Acotyledones.

Sporid'ium (σπορα seed, ειδος resemblance). The immediately enveloping membranous case to the sporules in certain Acotyledones; such cases themselves being included in a general one, or *ascus*. The term has also been applied to certain spore-like granules; and used synonymously with Spora and Sporulum; fig. 164.

164

Sporido'chia, *Sporido'chium* (σπορα a seed, δοχος capacious). Used synonymously with Acceptaculum, Stroma, Podetium, among certain groups of Acotyledones.

Sporocar'pium (σπορα a seed, καρπος fruit). Used synonymously with Apothecium, Ascus, and other forms and conditions of the spore-cases of Acotyledones.

Sporocla'dium (σπορα a seed, κλαδος a branch). A branch bearing the reproductive bodies of certain Algæ.

Sporocys'ta (σπορα a seed, κυστις a bladder). The sporocarpium of Algæ.

Sporoder'mis (σπορα a seed, δερμα skin). The integument or skin of a spore.

Sporo'phorum (σπορος a seed, φερω to bear). Synonyme for Trophospermum.

Sporophyl'lum (σπορος a seed. φυλλον a leaf). A sub-division of the thallus in Algæ bearing the fructification.

Sporota'mium (σπορος a seed, ταμειον a store-house), Synonyme for Receptaculum, when applied to the part under the disk of the shield of a Lichen.

SPO'RULE, *SPO'RULA* (diminutive for spora). A seed-like reproductive body in Acotyledones; synonymous with Spore.

SPORULIF'ERUS, SPORULI'GERUS (*SPORULA*, *FERO* and *GERO* to bear). Terms applied to the investing membrane immediately containing the sporules.

SPOTTED. When a colour is disposed in small spots on a ground of a different colour.

SPREADING. Where the tendency outwards, or bending from an axis, is gradual.

SPUMES'CENT, *SPUMES'CENS* (foaming). Having the appearance of foam or froth.

SPUR. A tubular expansion of some part more or less foliaceous; but especially among the floral whorls.

SPU'RIOUS, *SPU'RIUS* (counterfeit). Synonyme for False.

SQUA'MA (a fish-scale). A bract of the involucrum in Compositæ.

SQUAMA'TIO (*SQUAMA* a scale). When leaves are reduced to scale-like appendages, and are disposed in the form of a rosette at the end of a branch, the development of whose axis has been checked.

SQUAMA'TUS (*SQUAMA* a scale). Synonyme for Squamosus.

SQUAMEL'LA (diminutive for *SQUAMA*). A small scale-like bract, frequent on the receptacle of Compositæ.

SQUAMELLIF'ERUS (*SQUAMELLA*, and *FERO* to bear). Furnished with squamellæ.

SQUAMEL'LULA (diminutive from *SQUAMELLA*). A sub-division of the limb of the pappus of Compositæ. Scale-like appendages within the tube of certain corollæ.

SQUAMIFLO'RUS (*SQUAMA* a scale, *FLOS* a flower). A perianth of one or more scale-like bracts, to which the sexual organs are attached, but not disposed in a circle round an axis, as in Coniferæ.

SQUAMIFOR'MIS (*SQUAMA* a scale, *FORMA* shape). Shaped like a scale.

SQUAMO'SE, *SQUAMO'SUS* (scaly). Covered with scales; or composed of scale-like appendages.

SQUA'MULA (diminutive from *SQUAMA* a scale). Synonyme for Glumella. Synonyme for Squamella.

SQUAMULIFOR'MIS (*SQUAMULA*, and *FORMA* shape). Shaped like a small scale.

SQUAMULO'SUS (*SQUAMA* a scale). Covered with small scales.

SQUARRO'SE, *SQUARRO'SUS* (rough and scurfy). Where appendages diverge at a large angle from the axis, or the plane to which they are attached.

SQUARRO'SO-DENTA'TUS. When the teeth on the margin of a leaf are bent aside from the plane of its lamina.

SQUARRO'SO-LACINIA'TUS, *SQUARRO'SO-PINNATIF'IDUS*. Where the incisions of laciniate and of pinnatifid leaves are squarrosely disposed.

SQUARRULO'SUS (diminutive from *SQUARROSUS* rough). Slightly squarrose.

STACKHOU'SIACEÆ, STACKHOU'SIEÆ (from the genus Stackhousia). A natural order of Dicotyledones.

STA'CHYS (*σταχυς* an ear of corn). Synonyme for Spica.

STALK. Synonyme for Stem; and also for Petiole.

STALKLET. Synonyme for "secondary petiole," or that which supports a leaflet.

STAMEN. A floral organ containing the pollen.

STAMINA'LIS (*STAMEN* a stamen). Belonging to, or bearing relation to, stamens.

STAMINA'RIS (*STAMEN* a stamen). When a double flower is produced by the transformation of stamens into petals. Synonyme for Staminalis.

STAMI'NEAL, *STAMI'NEUS* (*STAMEN* a stamen). Having some marked reference to the stamens. As where the stamens are very prominent; or where perfect, and the corolla wanting.

STAMINI'DIUM (*STAMEN* a stamen) Organs in some cryptogamous plants, which have been considered analogous to the anthers of Phanerogamic species.

STAMINIF'ERUS, *STAMINI'GERUS* (*STAMEN* a stamen, *FERO* and *GERO* to bear). Bearing or supporting stamens.

STAMINO'DIUM (*στημων* a stamen, *ειδος* resemblance). An abortive stamen; or at least an organ bearing a resemblance to an abortive stamen.

STAMINO'SUS (*STAMEN* a stamen). Where the stamens, from size or elongation, form a marked feature in a flower.

STANDARD. See Papilionaceous.

STAPHYLLEA'CEÆ (from the genus Staphyllea). The Bladder-Nut tribe. A natural order of Dicotyledones.

STARRY. Synonyme for Stellate.

STARVED. When any part is less fully developed in a plant than it is in most other plants that are closely allied to it.

STATI'CEÆ (from the genus Statice). Synonyme for Plumbaginaceæ.

STATION, *STA'TIO*. Any locality within the "habitation" ascribed to a plant, in which the conditions necessary to its growth are established; viz. particular soil, amount of heat, moisture, &c.

STATOSPER'MUS (στατος firm, σπερμος seed). When a seed is straight or erect within the pericarp.

STAUROPHYL'LUS (σταυρος a cross, φυλλον a leaf). Synonyme for Cruciatus.

STEL'LATE, *STELLA'TUS* (full of stars). When several similar parts are disposed in a radiating manner round a centre.

STELLA'TÆ (from the stellate appearance of the flowers). The Madder tribe. A natural order of Dicotyledones.

STELLA'TO-PILO'SUS (*STELLATUS* and *PILOSUS*). When the pubescence is stellate.

STELLIF'ERUS, *STELLIFO'RMIS*, *STELLI'GERUS*, *STELLULA'TUS* (*STELLA* a star, *FERO* and *GERO* to bear, *FORMA* shape). Synonymes for Stellatus.

STEL'LULA (diminutive from *STELLA* a star). Synonyme for Rosella. The foliaceous whorl which invests the supposed anthers in mosses.

STEM. The ascending axis of a plant from which leaves, flowers, and fruit are developed.

STEM-CLASPING. See Amplexicaul.

STEMLESS. Where the stem is so little developed as to seem to be wanting.

STENOCAR'PUS (στενος narrow, καρπος fruit). Where the fruit is remarkably straight. So also with several other combinations with στενος.

STEPHANODO'PHYTUM (στεφανηδον shaped like a crown, φυτον a plant). Plant bearing a stephanoum.

STEPHA'NOUM (στεφανος a crown). Synonyme for Cremocarpium and Cypsela.

STEPHOCAR'PUS (στεφος a crown, καρπος fruit). Where a plant has its fruit arranged so as to resemble a crown.

STERCULIA'CEÆ (from the genus Sterculia). A natural order of Dicotyledones.

STERIG'MA, *STERIG'MUM* (στηριγμα a prop). Synonymes for Carcerulus. An elevated (more or less foliaceous) ridge, proceeding down the stem below the attachment of a decurrent leaf.

STER'ILE, *STER'ILIS* (barren). Where the fruit, or the pollen, is not perfected. Sometimes applied where there is only apparently such defect, but not a real one.

STICHID'IUM (στιχιδιον a little bladder). A case-like receptacle for the spores of some Algæ.

STICHOCAR'PUS (στῖχη row of any thing, καρπος fruit). Where the fruit is disposed along a spiral line.

STICTOPET'ALUS (στικτος dotted, πεταλον a petal). Where the petals are covered with glandular points.

STIGMA, *STIG'MA* (στιγμα a point). Exposed cellular tissue, free from epidermis, at one part (generally at the summit of the style or ovarium) of a carpel, where the fertilizing influence of the pollen is conveyed to the ovules. The term has been applied to a little mammillated point on the sporules of Equisetaceæ; and to a caducous point on the summit of the columella in Musci; and also to terminating points bearing the fructification of some Fungi.

STIGMA'TICUS (*STIGMA* the stigma). Belonging to the stigma.

STIGMATIFOR'MIS (*STIGMA* the stigma, *FORMA* shape). Shaped like a stigma.

STIGMATOI'DEUS (στιγμα the stigma, ειδος resemblance). Synonyme for Stigmatiformis.

STIGMATO'PHORUS (στιγμα the stigma, φερο to bear). The portions of the style which bear the stigma.

STIGMATOSTE'MON, *STIGMATOSTE'MONIS*, (στιγμα a stigma, στημων a stamen). Where the stamens cohere to the stigma.

STIG'MULA (*STIGMA* a stigma). Each of the several divisions of such stigmas as possess any.

STILAGINA'CEÆ, STILAGI'NEÆ, from the genus Stilago). A natural order of Dicotyledones.

STILBA'CEÆ, STILBI'NEÆ (from the genus Stilbe). A natural order of Dicotyledones.

STIM'ULANS (pricking). Stinging; see Sting.

STIMULO'SUS (*STIMULUS* a sting). When a surface is covered with stings.

STIM'ULUS. A sting.

STING. A sharp, somewhat stiff hair, seated on a gland which secretes an acrid fluid. When the skin of the human body is penetrated by the hair, and the fluid injected, this produces acute pain.

STIPEL'LA (diminutive from *STIPULA*). A minute stipule on a partial petiole of certain compound leaves.

Stipella'tus (from *stipella*), When partial petioles are furnished with stipellæ.

Stipel'lus (diminutive from *stipes*), Has been used synonymously with Filamentum.

Stipes (the trunk of a tree). Applied to certain very distinct descriptions of supports or props: — as the trunks of arborescent Monocotyledones and Filices; the stalk or petiole of the fronds of Filices; the support of the pileus in certain Fungi; a short stalk beneath the florets of certain Compositæ.

Stipif'erus (*stipes* and *fero* to bear). Where the receptacle of certain Compositæ carries a small stalk under each floret.

Stipifor'mis (*stipes*, and *forma* shape). Having the external appearance, but not the true character, of the stipes of endogenous trees.

Sti'pitate, *Stipita'tus* (from *stipes*). Furnished with a stalk-like support.

Stip'ticus (*stypticus* astringent). Possessing an astringent taste.

Stipula'ceus (from *stipula*). Possessing large stipules. Also, formed of scales which are degenerate stipules. Also, enveloped in stipules, which expand and enlarge as the leaf developes.

Stipula'ris (belonging to a *stipula*). Resulting from some peculiar modification of the stipule. Used, also, synonymously with *Stipulaceus*, in regard to plants furnished with unusually large stipules.

Stipula'tio (from *stipula*). Having relation to stipules.

Stipula'tus (from *stipula*). Furnished with stipules. Used also synonymously with Stipulaceus and Stipularis.

Sti'pule, *Sti'pula* (stubble). A foliaceous appendage, on each side the base of certain petioles, often laminated and membranous; but in some cases becoming a gland, a tendril, &c.

Stipulea'nus (from *stipula*). Resulting from the transformation of a stipule.

Stipulif'erus (*stipula*, and *fero* to bear). Supporting stipules.

Stipulo'sus (from *stipula*). Having very large stipules.

Stirpa'lis (from *stirps* the stock or stem of a tree). Growing on the stem.

Stirps (a kindred). A race.

Stock. Synonyme for a race. A plant to which a graft has been applied.

Stole, *Sto'lo* (a shoot). A lax trailing branch given off at the summit of the root, and taking root at intervals, whence fresh buds are developed.

Stolonif'erus (*stolo*, and *fero* to bear). Producing many stoles.

Sto'mate, *Sto'ma* (στομα the mouth). A very minute opening in the epidermis, between cells of a peculiar shape (generally reniform) distinct from that of the other cells. *Stoma* has also been employed synonymously with Epiphragma and Ostiolum.

Stomatif'erus (*stoma* and *fero* to bear). Furnished with stomates.

Stoma'tium. Synonyme for Stoma.

Stone. The hardened bony Endocarp of Drupes.

Stool. A plant from which "layers" are propagated, by bending its branches so that they may be inserted into and take root in the soil.

Stragling. Synonyme for "Divaricate."

Stra'gulum (a covering). Synonyme for Palea in grasses.

Straight. Without decided flexure of any kind.

Straight-ribbed. Either when the veins or nerves given off from the midrib of a dicotyledonous leaf are straight; or, when all the nerves of a monocotyledonous leaf run nearly straight from base to apex.

Straight-veined. Synonymous with straight-ribbed, in the case of monocotyledonous leaves.

Stra'men. Straw.

Strami'neus (belonging to straw). Straw-coloured. Yellow with a slight admixture of grey.

Strangula'ted. Irregularly contracted at intervals.

Strap-shaped. Linear, and in length somewhere about six times its own breadth; fig. 165.

165

Stra'tum (any thing strewed out). A layer or lamina composed of any kind of tissue.

Straw. The peculiar jointed stem of grasses.

Streak. A straight line formed by a vein, by colour, by indentation, &c.

Streaked. See Striated.

Strep'to-car'pus (στρεπτος twisted, καρπος fruit). Where

the fruit is marked by spirally arranged stripes. So of other combinations with strepto-.

STRI'A (a groove or furrow). A streak.

STRIA'TED, *STRIA'TUS* (channelled). Marked with striæ.

STRIC'TUS (close, narrow). Perfectly "straight."

STRI'GA (a row or ridge). A small straight hair-like scale.

STRI'GILIS (from *STRIGILIS* a curry-comb). Synonyme for Strigosus.

STRIGO'SE, *STRIGO'SUS* (from *STRIGA*). Covered with strigæ. Synonyme for "Hispid."

STRIPED. Marked with coloured "streaks."

STROBILA'CEUS, *STROBILIF'ERUS* (from *STROBILUS*, and *FERO* to bear). When flowers, furnished with large bracts, are so arranged as to give the inflorescence the appearance of a strobilus.

STROBILINUS (from *STROBILUS*). Either growing on a cone, or having the general aspect of a cone.

STRO'BILUS. A Cone.

STRO'MA (στρωμα a covering). Either, generally, the part of acotyledonous plants which bears or encloses the fructification; or, more restrictedly, limited to the fleshy thallus of certain Fungi, in which the perithecia are immersed.

STROMBUS-SHAPED, *STROMBULIF'ERUS*, *STROMBULIFOR'MIS* (*STROMBUS* a shell, spirally twisted like a screw, *FERO* to bear, *FORMA* shape). When the fruit is spirally twisted like a cork-screw; fig. 166.

166

STRO'PHES (στροφις a spiral). Any of the various spirals exhibited by the disposition of leaves round an axis.

STROPHIO'LE, *STROPHI'OLA*, *STROPHI'OLUS* (*STROPHIOLUM* a little chaplet). Synonyme for Caruncula.

STRUC'TURE, *STRUCTU'RA* (a building). The peculiar manner in which the several organs, elementary or compound, are disposed in plants.

STRU'MA (a wen). An apophysis which is restricted to one side of the base of a moss theca. A swelling at the point where the petiole is connected with the limb, in certain leaves.

STRUMI'FERUS, *STRUMO'SUS* (*STRUMA* a wen, *FERO* to bear). Furnished with a struma.

STRUMIFOR'MIS (*STRUMA* a wen, *FORMA* shape). Having the general appearance of a struma.

Strumulo'sus (diminutive of *strumosus* having a wen). Furnished with a small struma.

Stu'pa (tow). A tuft or mass of hair or fine filament matted together.

Stu'peus, Stup'peus (made of tow). Synonymes for Stuposus.

Stu'pose, *Stupo'sus* (*stupa* tow). Composed of matted filaments.

Sty'gius (infernal). Said of plants which grow in foul waters; in allusion to the Styx.

Styla'tus, Stylo'sus (from *stylus* the style). Where the style is remarkable for length; or by its persisting on the fruit.

Style, *Sty'lus* (a style or pin). A support frequently interposed between the stigma and ovary, and originating in the lengthening out of a part of the latter.

Stylidia'ceæ, Styli'deæ (from the genus Stylidium). A natural order of Dicotyledones.

Stylifo'rmis (*stylus* a style, *forma* shape). Resembling a style in its general aspect.

Styli'nus (*stylus* a style). Belonging to a style.

Stylis'cus (*stylus* a style). A chord of peculiar tissue, which descends from the stigma within the style, down to the ovary.

Stylo'deus (*stylus* a style). Furnished with a style.

Stylopo'dium (στυλος a column, and πους a foot). A fleshy disk at the base of each of the styles of the Umbelliferæ; fig. 167, *s*.

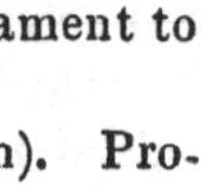

s.

167

Styloste'gium (στυλος a style, στεγη a roof) A peculiar form of cucullus surrounding the style.

Styloste'mon (στυλος, the style, and στημων a stamen). An epigynous stamen, originating in adhesion of the filament to the style.

Styloste'mus (στυλος a style, στημων a stamen). Proposed synonyme for Hermaphroditus.

Sty'lus. The style. Also the Ostiolum of certain Fungi.

Styra'ceæ, Styraci'næ, Styraci'neæ (from the genus Styrax). Either an order of Dicotyledones, or a sub-order of Ebenaceæ.

Styrido'phytus (σταυρος a cross, φυτον a plant). Synonyme for *Cruciformis*, applied to petals.

Sub (about or near to). Somewhat. When compounded

with any botanical term, implies a near approach to the condition which that term more precisely designates; thus, *Sub-acaulis*, when the stem is scarcely apparent, &c.

SUB-APICULA'RIS (*SUB* about, *APICULA* a sharp point). When the summit of a stem is a little prolonged, without branch or leaf, beyond a spike. The term has also been extended to panicles similarly circumstanced.

SUB-DIFFOR'MIS (*SUB* somewhat, and *DIFFORMIS*). Possessing a slight degree of irregularity.

SUBERO'SUS (*SUBER* cork). Of the nature of cork; or having some general resemblance to cork.

SUBLA'TUS (lifted up). When the ovary either has, or seems to have, a support.

SUBMER'GED, *SUBMERSI'BILIS*, *SUBMER'SUS* (sunk or drowned). Growing entirely beneath the surface of water.

SUB-NI'GER. Synonyme for Nigricans.

SUB-SERRA'TUS. Synonyme for Serrulatus.

SUB-SPE'CIES. Applied to some particular form, which may be considered as a more or less permanent variety of some species, rather than as characterizing a distinct species.

SUBTERRA'NEAN, *SUBTERRA'NEUS.* Synonyme for Hypogæan.

SUB'ULATE, *SUBULA'TUS*, *SUBULIFOR'MIS* (*SUBULA* a cobbler's awl, *FORMA* shape). Synonyme for Awl-shaped.

SUBULIF'ERUS (*SUBULA* an awl, *FERO* to bear). Furnished with long awl-shaped spines.

SUCCINC'TUS (trussed up). Synonyme for Circinatus.

SUCCI'NEUS (belonging to amber). Of an amber colour. Yellow with a little grey.

SUCCI'SUS (lopped off). Appearing as if a part were cut off at the extremity.

SUCCO'SUS. See Succulentus.

SUC'CUBUS (*SUCCUMBO* to couch down). The stipular appendages of certain Hepaticæ.

SUC'CULENT, *SUCCULEN'TUS* (*succus* juice). When the cellular tissue is abundant and replete with juices.

SUCKER. A branch (*surculus*) originating on a subterranean portion of a stem, and rising above ground. Also a tubercular process (*haustorium*) on the stems of certain flowering parasites, by which they imbibe nourishment from the plants to which they attach themselves.

SUFFRUTES'CENT *SUFFRUTES'CENS*, *SUFFRUTICO'SUS* (from *SUFFRUTEX*). Possessing the character of an under-shrub.

SUFFRU'TEX (*SUB* under, *FRUTEX* a shrub). An under-shrub.

SUFFUL'TUS (under-propped). When some part is so seated beneath another, as to appear as if it somewhat supported or propped it up.

SUL'CATE, *SULCA'TUS* (ranged in furrows). Marked by depressed parallel lines.

SUL'CUS (a furrow). Synonyme for Lamella in some Fungi.

SULPHU'REOUS, *SULPHU'REUS* (belonging to sulphur). A pale tint of pure yellow.

SUPER-AXIL'LARY, *SUPER-AXILLA'RIS*. See Supra-axillary.

SUPER-COMPOS'ITUS (*SUPER* more than). More than simply compound.

SUPERFICIA'RIUS (superficial). Upon the surface of an organ.

SUPER'FLUA. An order in the class Syngenesia of the Linnean system; containing plants whose capitula have the florets of the disk hermaphrodite, and those of the ray female.

SUPI'NUS (lying on the back, the face upward). Synonyme for Procumbens.

SUPRA-AXIL'LARY, *SUPRA-AXILLA'RIS* (*SUPRA* above, *AXILLA*, see Axil). Somewhat higher than the axil.

SUPRA-DECOM'POUND, *SUPRA-DECOMPO'SITUS* (*SUPRA* more than, *DECOMPOSITUS*, see "Decompound"). Where the extent to which a leaf is "Decompounded" cannot be definitely remarked.

SUPRA-FOLIA'CEUS (*SUPRA* above, *FOLIUM* a leaf). Placed higher upon a branch than a particular leaf.

SUPRA-FO'LIUS (*SUPRA* upon, *FOLIUM* leaf). Growing upon a leaf.

SURCULI'GERUS (*SURCULUS* a sucker, *GERO* to bear). Producing a sucker, or assuming the appearance of a sucker.

SUR'CULUS. A sucker. A young prostrate stem in some Mosses.

SURIANA'CEÆ (from the genus Suriana). A natural order of Dicotyledones.

SUR'SUM (upward). Directed upwards, and forwards.

SUSPEN'DED, *SUSPEN'SUS* (hung up). Attached somewhere between base and apex; and sometimes restricted to cases where the apex is directed downwards.

SUSPEN'SOR (*SUSPENSUS* hanged up). A very delicate cellular chord by which some Embryos appear to be attached to the apex of the nucleus.

Su'tural, *Sutura'lis* (*sutura* see Suture). Bearing some relation to the suture.

Sutura'rius (*sutura*, see Suture). Possessing a suture.

Su'ture, *Sutu'ra* (a seam). The plane of junction between contiguous parts, frequently indicated superficially by a line, either elevated or depressed.

Swim'ming. Used vaguely for aquatics, which either float on the surface, or have their leaves floating. More restrictedly applied to aquatics which are wholly immersed, and also free from attachment to the bottom.

Sword-shaped. A lamina with the edges sharp nearly parallel, but somewhat tapering from the base to the apex, which is acute; fig. 168.

168

Sychnocar'pous, *Sychnocar'pus* (συχνος frequent, αρποκς fruit). Where a plant produces fruit many times without perishing; as in the case of trees, shrubs, and perennials.

Syco'nium, *Sy'conus* (σῦκον a fig). An aggregate fruit where many flowers have been developed upon a fleshy receptacle, which is either a flattened disk, or forms a nearly closed cavity, as in the Fig; fig. 169.

169

Sygolli'phytum (συγκολλαω to fasten together). A plant where the perianth becomes combined with the pericarp.

Sylvat'icus, *Sylves'tris* (belonging to a wood; bred in the country). A plant whose natural habitat is a wood or copse.

Symmetran'thus (συμμετρια symmetry, ανθος a flower). When the perianth is separable by a line through the centre into two parts which are exactly alike.

Symmetrocar'pus (συμμετρια symmetry, καρπος fruit). When the pericarp is separable, by a plane through the axis, into two parts which are exactly alike.

Sym'metry, *Symme'tria*. When parts are so disposed round a centre, that a line or plane through that centre can separate them into two divisions, in each of which the parts are similarly disposed.

Sympeta'licus (συν together, πεταλον a petal). When stamens are combined into an androphorum, and, by adhesion to the petals, give to a strictly polypetalous flower the appearance of being monopetalous.

Symploci'neæ (from the genus Symplocos). Synonyme for Styraceæ.

SYMPHYANTHE'RUS (συμφυω to combine, ανθηρ an anther). Synonyme for Synantherus and Syngenesius.

SYMPHYOSTE'MON (συμφυω to combine, στημων a stamen). Synonyme for Monadelphus.

SYMPHY'SIA, *SYM'PHYSIS* (συμφυσις a concretion). A growing together, or uniting of parts usually distinct.

SYMPHYTANTHE'RUS. See Symphyantherus.

SYMPHYTO'GYNUS, *SYMPHYTOTHE'LUS* (συμφυω to combine, γυνη a woman; θηλη a nipple). Indicating adhesion between the calyx and pistil; where the ovary is more or less inferior.

SYMPLO'CIUM, *SYMPLO'KIUM* (συν together, πλεκω to bind). Synonyme for the Annulus to the Thecæ of Ferns.

SYNAN'THEREÆ (συν together, ανθηρ an anther). Synonyme for Compositæ.

SYNAN'THEROUS, *SYNANTHER'ICUS*, *SYNANTHE'RUS* (συν together, ανθηρ an anther). Synonyme for Syngenesious.

SYNARMO'PHYTUS (συναρμοσις conjunction, φυτον a plant). Synonyme for Gynandrous.

SYNANTHRO'PHYTUM (συν together, αθροιζω to congregate, φυτον a plant). A plant whose fruit is composed of many carpels aggregated together.

SYNCAR'PIUM, *SYNCAR'PUM* (συν together, καρπος fruit). A fruit composed of several carpels, which become more or less fleshy and cohere together. Applied also where the carpels are combined by their floral envelopes becoming fleshy.

SYNCAR'POUS, *SYNCAR'PUS* (συν together, καρπος fruit). When a plant bears fruit composed of cohering carpels.

SYNCOTYLEDO'NEUS (συν together, κοτυληδων a cotyledon). When the cotyledons of Dicotyledonous plants so cohere together as to form a single undivided mass.

SYNE'DRUS (συνεδρος a sitter by). Growing on the angle of a stem.

SYNE'MA (συν together, νημα a thread). The portion of the Gynostemium corresponding to the position of the combined filaments.

SYNGENE'SIA (συν together, γενεσις origin). The nineteenth Class of the artificial system of Linneus; originally including all plants whose anthers cohere; but now restricted to plants belonging to the vast natural order of Compositæ.

SYNGENE'SICUS. Belonging to Syngenesia.

SYNGENE'SIOUS, *SYNGE'NESUS* (συν together, γενεσις origin). Where the stamens cohere by their anthers.

SYNOCHO'RION (συν together, χοριον membrane surrounding the fœtus). Synonyme for Carcerulus.

SYNORHI'ZUS (συν together, ριζα a root). When the point of the radicle, in the embryo, adheres to the perisperm.

SYNSTIGMA'TICUS (συν together, στιγμα the stigma). When a pollen mass, in the Orchideæ, is furnished with a retinaculum, by which it adheres to the stigma.

SYNTRO'PHICUS (συν together, τρεφω to nourish). Synonyme for Epiphyticus.

SYNZY'GIA (συν together, ζυγοω to yoke). The point of junction, above the radicle, where opposite cotyledons meet in Dicotyledonous plants.

SYSTELLO'PHYTUM (συστελλω to confine, φυτον a plant). When a persistent calyx appears to form part of the fruit.

SYSTEM, *SYSTE'MA* (συν together, ιστημι to stand). An arrangement of natural objects according to prescribed rules.

SYSTEMA'TIC, *SYSTEMA'TICUS*. Belonging to a system. Also, when the species of a genus are distinguished from those of allied genera by a single character.

SYS'TYLUS (συν together, στυλος the style). Where several styles cohere so as to form a single column.

TABACI'NUS (*TABACUM* tobacco). Tobacco-coloured. Grey with some binary compound of red with a little yellow.

TA'BES (a rotting away). A disease which produces a gradual decay.

TABES'CENS (wasting away). Synonyme for Marcescens.

TA'BULA (a table). Synonyme for Pileus in some Fungi.

TABULA'TUS (boarded). Consisting of superimposed layers.

TACCA'CEÆ, TAC'CEÆ (from the genus Tacca). A natural order of Monocotyledones.

TÆNIA'NUS (*TÆNIA* a ribband, and hence a tapeworm). Tapeworm-shaped.

TAIL. Any long, flexible, and terminal appendage to various parts. Often used as a synonyme for Petiole, and also for Peduncle, in common parlance.

TAIL-POINTED. Terminated by a much elongated and weak point.

TALA'RA (*TALARIUM* the winged shoe of Mercury). Synonyme for an Ala in the Papilionaceous corolla.

Ta'lea. A slip or cutting, made for the purpose of propagating the plant.

Tall. Exceeding the usual height attained by other species of the same genus.

Tamarica'ceæ, Tamarisci'neæ (from the genus Tamarix). The Tamarisk tribe, A natural order of Dicotyledones.

Taper. Long, slender, and where the sections perpendicular to the axis are circular; the form being really cylindrical, or nearly so.

Ta'pering. Where there is a very gradual diminution in the diameters of transverse sections of an otherwise "taper" form.

Ta'per-pointed. Synonyme for Acuminate.

Tapeworm-shaped. With the general appearance of a tapeworm; long, linear, with contractions at intervals.

Tap-root. An unbranched tapering mass round the descending axis, from which fibres are given off in more or less abundance.

Tarta'reous, *Tarta'reus* (*tar'tarum* the tartar-crust in wine vessels). Where a surface is rough and crumbling, as in many Lichens.

Taw'ny. Of a dull yellowish tint, obtained by mixing orange yellow with grey.

Taxa'ceæ, Taxi'næ (from the genus Taxus). A natural order of Dicotyledones.

Taxiformis (*taxus* the Yew, *forma* shape). Arranged like leaves of the Yew, in a distichous manner.

Taxo'logy, Taxo'nomy, *Taxolo'gia*, *Taxono'mia* (ταξις order λογος a word, νομος a law). That department of Botany which embraces the classification of plants.

Tear-shaped. A slight modification of pear-shaped, applied to solids, without contraction or curvature inwards; fig. 170.

170

Teg'men (a covering). See Secundine. Synonyme for Gluma in grasses.

Tegmen'tum (a covering). The outer scales of a leaf-bud.

Tegmina'tus (*tegmen* a covering). When the nucleus is invested by a tegmen.

Tegumen'tum (a covering). Synonyme for Indusium in Ferns. Synonyme for Spermodermium.

Te'la (a web of cloth). Elementary tissue.

Teleian'thus (τελειος perfect, ανθος a flower). Synonyme for Hermaphroditus.

TEN'DRIL. A modified condition of some append. age to the axis of vegetation, when it assumes the form of a chord, twisting round contiguous objects for support; fig. 171.

TENUIFO'LIUS (*TENUIS* slender, *FOLIUM* a leaf). When the subdivisions of a leaf are linear and slender.

TE'PALUM (formed by analogy from *PETALUM* and *SEPALUM*). The subordinate parts of a perianth when these are not referred to either calyx or corolla.

TEPHRO'SIUS (τεφρος ash-colour). Of an ash-grey colour. Grey much diluted.

TERATOL'OGY (τερας a prodigy, λογος a discourse). Synonyme for Morphology.

TERCI'NE, *TERCI'NA* (*TER* thrice). A membrane between the secundine and nucleus; but not generally allowed to be any more than some layer of the regular envelopes to the later.

TEREBINTA'CEÆ, TEREBINTHA'CEÆ (from the Pistacia Terebinthus). A natural order of Dicotyledones.

TERE'DO (a boring animal). Disease produced by the perforations of insects.

TE'RETE, *TE'RES* (long and round). Nearly cylindrical, but somewhat tapering into a very elongated cone.

TERGEM'INATE, *TERGEM'INUS*, *TERGEMINA'TUS* (*TER* thrice, *GEMINUS* double). When three pairs of leaflets are attached by secondary petioles to a common petiole; fig. 172.

TERGISPER'MUS (*TERGUM* a back, σπερμα a seed). Where the fructification is borne on the back of the frond, as in some Ferns. *DORSIF'ERUS* is the better term.

TER'GUM. The Back.

TER'MINAL, *TERMINA'LIS* (*TERMINUS* an end). Situate at the extremity of some part.

TERMINOL'OGY (*TERMINUS* a term or word used technically, λογος a discourse). Synonyme for Glossology.

TERNA'TE, *TERNA'TUS* (*TERNUS* three and three). Arranged by threes, about the same part.

TERNA'TO-PINNA'TUS. When three secondary petioles, attached to a common petiole, bear leaflets arranged in a pinnate manner.

TERNSTRŒMIA'CEÆ, TERNSTRŒM'IEÆ (from the genus Ternstrœmia). A natural order of Dicotyledones.

TER'NUS. See Ternate.

TERRA'NEUS TERRES'TRIS (living on the earth). Growing above the surface, on dry land.

TESSELLA'TED, *TESSELLA'TUS* (in checquer work). When colours are so blended in nearly square or oblong patches, as to represent an appearance of chequer work.

TESSULA'RIS (*TESSERA* a die). Approaching the form of a cube.

TES'TA (a pot). The outer coat of the Spermoderm.

TESTA'CEOUS, *TESTA'CEUS* (made of brick or tile). Brownish yellow. Orange yellow with much grey.

TESTIC'ULA, TESTIC'ULATE, *TESTICULA'TUS* (from *TESTICULUS*). Solid and ovate.

TESTIC'ULUS, *TES'TIS*. Synonyme for *ANTHERA*.

TE'TER (stinking). Giving out a fœtid odour.

TETRACA'MARUS (τετρα four, καμαρα a vault). A fruit composed of four *CAMARÆ*.

TETRACHÆ'NIUM (τετρα four, χαινω to open). A fruit formed by the separating of a single ovary into four nuts; as in the Labiatæ.

TETRACHOCAR'PIUM (τετρα four, καρπος fruit). A cluster of four spores in certain Algæ.

TETRACHO'TOMUS (τετραχως four ways, τεμνω to divide). When a cyme (in its restricted sense of fascicle) bears four lateral peduncles about the terminal flower.

TETRACOC'CUS (τετρα four, κοκκος a seed). A fruit composed of four *COCCA*.

TETRA'DYMUS (τετραδυμος four double). Where every alternate lamella of an Agaric is shorter than the two contiguous to it, and one complete lamella terminates a set of every four pairs of short and long. Also, where four cells or cases are combined.

TETRADYNA'MIA (τετρα four, δυναμις power). An artificial Class of the Linnean system, containing hexandrous flowers in which four of the stamens are longer than the other two; fig 173. It agrees with the natural order Cruciferæ.

173

TETRADY'NAMOUS, *TETRADY'NAMUS*, Where the stamens are arranged as in Tetradynamia.

TETRAFOLIA'TUS (τετρα four, *FOLIUM* a leaf). Synonyme for Bijugatus.

TETRAGONIA'CEÆ (from the genus Tetragonia). A natural order of Dicotyledones.

TETRA'GONOUS, *TETRAGO'NUS* (τετρα four, γωνια an angle). Having four angles, not very acute.

TETRAGY'NIA (τετρα four, γυνη a woman). An order to some of the classes of the artificial system of Linneus, characterized by the flowers having four distinct pistils, or four distinct styles on one pistil.

TETRA'GYNOUS, *TETRA'GYNUS*. Where the four carpels, or at least styles, are free, as in Tetragynia.

TETRAN'DRIA (τετρα four, ανηρ a man).

THECID'ION, *THECI'DIUM* (θηκη a box). Synonyme for Achenium.

THECI'GERUS (*THECA* and *GERO* to bear). Applied to the hymenium of Fungi, and the branches of such Musci as produce no setæ.

THELE'PHORUS (θηλη a nipple, φερο to bear). Covered with nipple-like prominences.

THICK, THICKENED. When the thickness of an organ is relatively greater, with respect to its size, than is usual in similar organs.

THORN. See Spine.

THREAD-SHAPED. See Filiform.

THREE-CORNERED, THREE-EDGED. Prismatic, with three sides plain (or somewhat curved), and the angles more or less acute.

THREE-RIBBED. When the midrib and a nerve on each side of it from near the base are much more strongly exposed than the other nerves of a leaf; fig. 174.

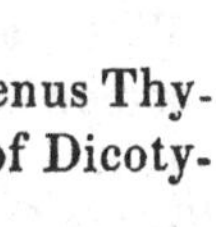

174

THRICE-DIGITATO-TERNATE. Synonyme for Triternate.

THROAT. The opening in the upper part of the tube of a monopetalous corolla.

THYMELA'CEÆ, THYME'LEÆ (from the exploded genus Thymelæa). The Mezereum tribe. A natural order of Dicotyledones.

THYRSE, *THYR'SUS* (the ancient thyrsus). A branched raceme, in which the middle branches are longer than those above or below them.

THYR'SULA (diminutive from Thyrsus). Synonyme for Verticillaster.

THYRSIF'ERUS, *THYRSIFLO'RUS* (*THURSUS*, *FERO* to bear, *FLOS* a flower). Where the inflorescence is a thyrsus.

TIGEL'LA (a latinized word from the French *TIGELLE*, diminutive of *Tige* a stem). The portion of the embryo between the radicle and cotyledons.

Tigella'tus (from *tigella*). When the tigella is well marked.

Tigellu'la (diminutive of *tigella*). A description of filament in the Truffle.

Tigellula'ris. Synonyme for Vascularis.

Tilia'ceæ (from the genus Tilia). The Linden tribe. A natural order of Dicotyledones.

Tincto'rius (*tinctus* dyed). Capable of serving as a dye.

Toise. A measure from about five to six feet.

Tomen'tose, *Tomento'sus.* Covered with *tomentum.*

Tomen'tum (flocks of wool). Pubescence, which consists of hair closely matted, very slightly rigid, and rather short.

Tongue-shaped. Long, fleshy, plano-convex, and obtuse.

Tooth. A small projection, generally resulting from an apparent jagging or incision of the margin of some laminated part.

Toothed. Furnished with teeth.

Top-shaped. Conical and somewhat contracted towards the apex of the cone, which is the point of attachment.

Torfa'ceus. Synonyme for Turfosus.

Torn. When marginal incisions are deep and irregular.

Toro'sus. Synonyme for Torulosus.

Torsi'vus (*tortus* twisted). Spirally twisted.

Tor'tilis (winding). With a tendency to twist; or capable of being twisted spirally.

Tor'tuous, *Tortuo'sus* (crooked). Bent irregularly in different directions.

Torulo'sus (*torulus* a ringlet peculiarly twisted). When a cylindrical body is swollen at intervals, somewhat resembling a chord with knots; fig. 175.

To'rus (a bed). The axis on which all the parts of the floral whorls within the calyx are seated.

Trabe'cula (a small beam). A connection, like a cross-bar, uniting contiguous teeth in some Mosses.

Trachæ'a, Trache'a (the wind-pipe). A minute cylindrical vessel (tapering at the extremities to points) of the elementary tissue, composed of membrane, with one or more spirally-twisted fibres lining the interior; fig. 176.

Trachen'chyma (*trachea*, and χευω to diffuse). Fibro-vascular tissue, composed of tracheæ, or of ducts with markings spirally arranged.

Trachycar'pus, *Trachysper'mus* (τραχυς rough, καρπος fruit, σπερμα seed). Used synonymously in some cases where the pericarp is rough with points.

Trajec'tilis (*trajectio* a displacing). When the connective completely separates the anther cells.

Tra'ma (yarn). The tissue of certain Fungi.

Trans'verse, *Transver'sus*, *Transversa'lis*. In a plane perpendicular to the axis, or longitudinal direction.

Trapezo'id, Trape'ziform, *Trapezoi'deus* *Trapezifor'mis* (τραπεζιον a trapezium). When the sides of some four-sided lamina (as the leaf) are unequal.

Tree. A woody plant with a trunk, or single stem rising above ground.

Tree-like. Having a general and miniature resemblance to a tree, but not being one.

Treman'draceæ, Treman'dreæ (from the genus Tremandra). A natural order of Dicotyledones.

Tremelloi'des *Tremello'sus*. Having the consistency and appearance of the genus Tremella.

Trewia'ceæ (from the genus Trewia). A natural order of Dicotyledones.

Triache'nium (*tres* three, and *achenium*). An inferior simple fruit, which on ripening separates into three parts. See Cremocarp.

Triadel'phous, *Triadel'phus* (τρεῖς three, αδελφος a brother). Where the stamens are collected into three distinct bundles, the filaments of those in the separate bundles cohering.

Triake'nium. See Triachenium.

Trian'dria (τρεις three, ανηρ a man). A class of the Linnean system, containing hermaphrodite flowers with three stamens.

Trian'drous, *Trian'der*, *Trian'drus* (τρεις three, ανηρ a man). Having three stamens.

Trian'gular, *Triangula'ris*, *Triangula'tus*. Either a plain surface approaching a triangle in shape; or a solid part whose transverse section approaches a triangle with acute angles; fig. 177.

177

Triangula'to-cunea'tus. Between *Triangulatus* and *Cuneatus* in shape.

TRIAN'THUS (τρεις three, ανθος a flower). When a peduncle bears three flowers.

TRIBE, *TRIBUS*. A group of genera subordinate to an order.

TRI'CA. A form of apothecium, where the surface is orbicular, and presents spirally-disposed and elevated lines.

TRICA'MERUS (τρεις three, and *CAMARA*). A fruit composed of three camara.

TRICEPH'ALUS (τρεις three, κεφαλη a head). When a fruit is composed of three carpels united below, but more or less distinct above. When capitula occur three and three together.

TRICHI'DIUM (τριχιον hair). Hair-like filaments bearing the spores of certain Fungi.

TRICHOCAR'PUS (τριχιον or θριξ hair, καρπος fruit). Where the fruit is covered with hair-like pubescence.

TRICHOCEPH'ALUS (θριξ hair, κεφαλη a head). Where flowers are collected in heads, which are surrounded with hair-like appendages.

TRICHO'DES (θριξ hair, ειδος resemblance). Resembling hair.

TRICHOLO'MA (θριξ hair, λωμα border). When an edge or border is furnished with hairs.

TRICHO'MA (θριξ hair). Hair like filaments composing the thallus of certain Algæ.

TRICHOMY'CES (θριξ hair, μυκης a fungus). An order or tribe of Fungi, characterized by their filamentous appearance.

TRICHOPH'ORUM (θριξ hair, φερω to bear). When the stipes of certain Fungi are formed by the union of filaments.

TRICHOPHYL'LUS (θριξ hair, φυλλον a leaf). Where a leaf is either hair-like, or terminates in a hair.

TRICHOT'OMUS (τριχη by threes, τεμνω to cut). When subdivisions of any part takes place by threes.

TRICOC'CÆ. A group composed of plants whose fruit is a *TRICOCCUS*.

TRICOC'CUS (τρεις three, κοκκος a seed). A fruit composed of three cocci. N. B. *Coccus* should have been given as a synonyme to *COCCUM*, in this Dictionary.

TRICO'LOR (of three colours). Where a flower exhibits three distinct colours.

TRICOTYLE'DONUS (τρεις three, κοτυληδων a seed leaf). A dicotyledonous plant, whose embryo is furnished with three cotyledons; or else one of the two is so deeply lobed as to appear double.

TRICUSPIDA'TUS (*TRES* three, *CUSPIS* a point). Any part furnished with three distinct points or projections.

TRIDENT-POINTED, *TRIDENTA'TUS* (*TRES* three, *DENTATUS* toothed). Having three divisions in the form of teeth; fig. 178.

178

TRIDIGITA'TO PINNA'TUS (*TRES* three, *DIGITATUS* and *PINNATUS*). Synonyme for Ternato-pinnatus.

TRIDIGITA'TUS (*TRES* three, and *DIGITATUS*). Synonyme for Ternatus.

TRIDU'US (*TRIDUUM* three days long). Lasting for three days.

TRI'DYMUS (τρεις three, δυναμος power). When, of three laminæ, in Agarics, ranging between two which extend to the margin, the middle one is larger than the other two.

TRIDY'NAMUS (τρεις three, δυναμος power). When three out of six stamens are longer than the other three.

TRIE'DER (τρεις three, εδρα a base). Synonyme for Triangularis.

TRIENNA'LIS, *TRIEN'NIS* (*TRIENNIUM* of three years duration). Lasting for three years.

TRIFA'RIUS (*TRIFARIUM* three manner of ways). Ranged in three rows.

TRI'FID, *TRI'FIDUS* (*TRES* three, *FISSUS* cleft). Where the incisions extend about half-way towards the base in a divided lamina. When any part is divided into three subordinate parts.

TRIFO'LIATE, *TRIFOLIA'TUS*, *TRIFO'LIUS* (*TRES* three, *FOLIUM* a leaf). Applied when leaflets are disposed in threes at the extremities of their petioles, whether in compound or decompound leaves.

TRIFOLIOLA'TUS (*TRES* three, *FOLIUM* a leaf). Synonyme for Ternatus.

TRIFOR'MIS (*TRES* three, *FORMA* shape). When the receptacle, in Compositæ, bears florets of three different shapes.

TRIFRONS (*TRES* three, *FRONS* a frond). When a Fern has fronds of three distinct forms.

TRIFURCA'TUS (*TRES* three, *FURCA* a fork). Divided towards the summit into three points.

TRI'GAMUS (τρεις three, γαμος a marriage). When the receptacle in Compositæ bears three kinds of florets differing with respect to sex.

TRI'GLANS (*TRES* three, *GLANS* an acorn). Where there are three *glandes* seated on the *cupula* in Cupuliferæ.

Trigonocar'pus. (τριγωνος a triangle, καρπος fruit). Where the fruit is three-sided, the angles distinctly marked.

Tri'gonus (a triangle). Either triangular, or when a transverse section is so.

Trigy'nia (τρεις three, γυνη a woman). An Order to some classes of the Linnean system, where flowers have either three pistils, or at least three distinct styles.

Trigy'nus. Where the pistils are arranged as in Trigynia.

Trihila'tus (*tres* three, *hilum* a speck. Having three openings.

Tri'jugus (*tres* three, *jugum* a yoke). Having three pairs of leaflets, in a pinnate leaf.

Trilat'eral, *Trilatera'lis* (*tres* three, *lateralis* belonging to a side). Prismatic, and three-sided.

Triloba'tus, *Tri'lobus* (*tres* three, *lobus* a lobe). Divided into three lobes.

Trilo'cular, *Trilocula'ris* (*tres* three, *locula* a cell). Divided interiorly into three cells.

Trinerva'tus, *Triner'vis* *Triner'vius* (*tres* three, *nervus* a nerve). When a leaf has three strongly defined nerves proceeding from the base.

Trinervula'tus (*tres* three, *nervulus* a little sinew). When three nerve-like vascular chords occur in a placenta.

Triœ'cia (τρεις three, οικος a house). An Order in the Linnean Class Polygamia where a species produces each of the three kinds of flowers on separate individuals.

Triœ'cious, *Triœ'cius*, *Triœ'cicus*, *Trioi'cus*. Having the flowers circumstanced as in Triœcia.

Triopercula'tus (*tres* three, *operculum* a lid). With three lids.

Triovula'tus (*tres* three, and *ovula* an ovule). When an ovary contains three ovules.

Tripar'ted. Synonyme for Tripartite.

Triparti'bilis (*tres* three, *pars* a part). Capable of separating by dehiscence into three parts.

Tripar'tite, *Triparti'tus*. (*tres* three, *pars* a part). Subdivided into three parts, much beyond the middle or nearly to the base.

Tripen'nate, *Tripenna'tus*. See Tripinnate.

Tripet'aloid, *Tripetaloi'deus* (τρεις three, πεταλον a petal, ειδος resemblance). When a six-leaved perianth, in Monocotyledones, has the three inner segments more highly coloured and petal-like than the three outer.

TRIPET'ALUS (τρεὶς three, πεταλον a petal). When a corolla consists of three petals.

TRIPHYL'LUS (τρεις three, φυλλον a leaf). Having some reference to a ternary disposition of the leaves, or the parts resulting from their metamorphosis.

TRIPIN'NATE, *TRIPINNA'TUS* (*TRES* three, *PENNA* a wing). Where the leaflets (or some of them) of a triply compound leaf are arranged in a pinnate manner; fig. 179. Compare with Bipinnate.

179

TRIPINNAT'IFID, *TRIPINNATIF'IDUS*. Where the divisions of a bipinnatifid leaf are cut in a pinnatifid manner.

TRIPPLE-NERVED, TRIPPLE-RIBBED, *TRIPLINERVA'TUS* (*TRIPLEX* tripple, *NERVUS* a nerve). When a prominent nerve branches off on either side of the midrib of a leaf near the base, the other nerves being comparatively small.

TRI'PLEX, *TRIP'LUS* (tripple). Exhibiting a three-fold division.

TRIPLICA'TO-GEMINA'TUS. Synonyme for Tergeminatus.

TRIPLICA'TO-NERVA'TUS. Synonyme for Tripli-nervatus.

TRIPLICA'TO-PENNA'TUS, or *PINNA'TUS*. Synonyme for Tripinnatus.

TRIPLINER'VIS, *TRIPLINER'VIUS*. Synonymes for Triplinervatus.

TRIP'TERUS (τρεις three, πτερον a wing). Furnished with three wings.

TRIQUE'TER, *TRIQUE'TRUS* (three-cornered). With three faces and edges.

TRIQUINA'TUS (*TRES* three, *QUINARIUS* containing five). When the lower divisions of a Bipinnatifid leaf are trifid.

TRISERIA'LIS, *TRISERIA'TUS* (*TRES* three, *SERIES* a series). Disposed in three rows.

TRISTA'CHYUS (τρεις three, σταχυς a spike). Where the peduncles bear three spikes.

TRIS'TICHOUS, *TRIS'TICHUS* (τρεὶς three, στιχος a row). Synonyme for Triserialis.

TRISTIGMA'TEUS (*TRES* three, and *STIGMA* the stigma). When either the ovary or a single style bears three stigmas.

TRIS'TIS (sad). Of a dingy or dull colour.

TRISULCA'TUS (*TRES* three, *SULCUS* a furrow). Marked by three depressed lines, or furrows.

TRITER'NATE, *TRITERNA'TUS* (*TRES* three, *TERNATUS*, see Ternate). Where the petiole becoming twice compound in a ternate manner, each partial petiole bears three leaflets; fig. 180.

TRIV'IAL, *TRIVIA'LIS*. The common or specific name given in addition to the generic name of a plant.

TROCHLEA'RIS (*TROCHLEA* a pulley). Pulley-shaped.

TROPÆO'LEÆ (from the genus Tropæolum). The Nasturtium tribe. An order of Dicotyledones.

TROPHOSPER'MIUM (τροφω to nourish, σπερμα the seed). Synonyme for Placenta.

TROP'ICAL, *TROP'ICUS*. Growing near or between the tropics. Flowers which expand in the morning, and close at night, during several successive days.

TRUMPET-SHAPED. Tubular and dilated at one end.

TRUN'CATE, *TRUNCA'TUS* (maimed). Terminating abruptly, as though it had been shortened by the removal of the extremity.

TRUNCATULA'RIS, *TRUNCULA'TUS* (diminutive of *truncatus*). Slightly truncate.

TRUNK, *TRUN'CUS* (a stump of a tree). Now restricted to the main stem (without its branches) of Dicotyledonous trees; but in its more extended signification was applied to all stems.

TRY'MA (τρυμα a hole). Drupaceous, superior, and the endocarp of two firmly combined valves.

TUBÆFOR'MIS, *TUBA'TUS* (*TUBA* a trumpet, *FORMA* shape). Trumpet-shaped.

TUBA'TUS. See Tubular.

TUBE, *TU'BUS*. The tubular portion formed by the cohesion of the subordinate parts composing a floral whorl.

TU'BER, *TU'BER* (an excrescence). A very fleshy, swollen, and subterranean rhizoma; of which the Potato is a familiar illustration.

TU'BERCLE, *TUBERCULUM* (a pimple). A small wart-like excrescence. Synonyme i a form of Apothecium.

TU'BERCLED, *TUBERCULA'TUS* (*TUBERCULUM* a pimple). Covered with tubercles.

TUBER'CULAR, *TUBERCULO'SUS* (*TUBERCULUM* a pimple). Having swollen appendages, or excrescences.

TUBERIF'ERUS (*TUBER* a swelling, *FERO* to bear). Bearing tubers or tuberous masses.

TU'BEROUS, *TUBERO'SUS* (*TUBER* an excrescence). Resembling a tuber, but not originating in an alteration of the stem.

TUBIFLO'RUS (*TUBUS* a tube, *FLOS* a flower). Where the tube of a monopetalous corolla is very long.

TUBIFOR'MIS (*TUBUS* a tube, *FORMA* shape). Resembling a tube.

TUBIL'LUS (diminutive of *TUBUS* a tube). Elongated cells of cellular tissue. The tube formed by the union of the filaments in Compositæ.

TU'BULAR, *TUBULA'RIS*, *TUBULO'SUS* (*TUBULATUS* hollowed like a pipe). Hollow and cylindrical.

TUBULIFLO'RUS (*TUBULUS* a little pipe, *FLOS* a flower). Bearing tubular florets.

TUBULIFOR'MIS (*TUBULUS* a little pipe, *FORMA* shape). Synonyme for Tubular.

TU'BULUS (a little pipe), *TU'BUS* (a tube). One of the pores which perforate the hymenium in certain Fungi.

TUITA'NS (*TUEOR* to defend). When leaves, during sleep, incline downwards, and appear, as it were, to protect the stem.

TULIPA'CEÆ (from the genus Tulipa). A synonyme for Liliaceæ.

TU'MIDUS (swollen). Synonyme for Inflatus.

TU'NICA (a tunic). A loose membranous skin investing some organ. Synonyme for Spermoderm. The Peridium of some Fungi.

TUNICA'TUS (coated). When invested with a Tunica.

TUR'BINATE, *TURBINA'TUS* (shaped like a top). Top-shaped.

TURBINIFOR'MIS (*TURBO* a top, *FORMA* shape). Synonyme for Turbinatus.

TURFA'CEUS, *TURFO'SUS*. Used as Torfaceus.

TUR'GIDUS (swollen). Thick, and as if swollen, but not inflated with air.

TU'RIO (a tendril). The early stage of a sucker, when invested by leaf scales.

TURIONIF'ERUS (*TURIO*, and *FERO* to bear). Throwing up turiones.

TURNED-INWARDS. Introrse.

TURNED-OUTWARDS. Extrorse.

TURNIP-SHAPED. Oblately-sphærical, and tapering below, as in a common Turnip.

101.

TURNERA'CEÆ (from the genus Turnera). A natural order of Dicotyledones.

TWIG-LIKE. Long, flexible, and wand-like.

TWIN. Synonyme for Gemminate.

TWIN-DIGITA'TO-PIN'NATE. Synonyme for Bi-digitato-pinnate.

TWI'NING. Twisting in spiral folds round a support.

TWO-EDGED. Longitudinally compressed, and presenting two sharp angles running parallel to the axis.

TYM'PANUM (a drum). A membrane closing the mouth of the theca in some Mosses.

TYPHA'CEÆ, TY'PHÆ, TYPHI'NÆ, TYPHOI'DEÆ (from the genus Typha). The Bulrush tribe. A natural order of Monocotyledones.

TYP'ICAL, *TYP'ICUS* (*TYPUS* a type). Especially presenting the chief characteristics of a particular group.

ULIGINA'RIUS, *ULIGINO'SUS* (marshy). Growing in marshy places.

ULMA'CEÆ (from the genus Ulmus). The Elm tribe. A natural order of Dicotyledones.

UL'NA (a man's arm). Estimated at about twenty-four inches.

ULNA'RIS. Of the length of an ulna.

U'LOTHRIX (ουλος frizzled, θρίξ hair). In hair-like crisped linear divisions.

UM'BEL, *UMBEL'LA* (a head of flowers, as in Fennel). A form of inflorescence in which all the pedicels start from the summit of the peduncle; fig. 181.

181

UMBELLA'TUS. When the inflorescence is in umbels, or approaches to such a disposition of the flowers.

UMBELLIF'ERUS (*UMBELLA* an umbel, *FERO* to bear). Bearing umbels. Assuming the form as an umbella.

UMBELLIFLO'RUS (*UMBELLA* an umbel, *FLOS* a flower). Synonyme for Umbellatus.

UMBELLIFOR'MIS (*UMBELLA* an umbel, *FORMA* shape). Synonyme for Umbelliferus.

UMBEL'LULA (diminutive of *UMBELLA*). Synonyme for "partial umbel." See "General."

UMBELLULA'TUS (*UMBELLA* an umbel). When the flowers are nearly disposed in the form of an umbel.

UMBELLULIF'ERUS (*UMBELLULA* a partial umbel, *FERO* to bear). With few flowers from the end of a common peduncle.

UM'BER. A dark Brown. Grey with a little red.

UMBIL'ICAL-CHORD (*umbilicus* the navel). Synonyme for "Funicular-chord."

UMBIL'ICATE, *Umbilica'tus* (*umbilicus* the navel). Having a depression in the centre. Also (*umbilicus* a boss) with an elevation in the centre. Synonyme for Peltate.

Umbili'cus (the navel). Synonyme for Hilum. Synonyme for Ostiolum; and, generally, either a depression or an elevation about the centre of a given surface.

UM'BO (a boss). Synonyme for Umbilicus when applied to a central elevation; fig. 182.

182

Umbona'tus. Furnished with an Umbo.

Umbonula'tus (diminutive from Umbonatus). When an Umbo is very small.

Umbraculifor'mis (*umbraculum* an umbrella, *forma* shape). Umbrella-shaped.

Umbra'culum (an umbrella). Having the general form of an umbrella.

Umbrati'colus (*umbra* a shade, *colo* to inhabit). Spontaneously vegetating in shady situations.

UMBRELLA-SHAPED. Hemispherical, or nearly so, elevated on a stipes, and with or without the appearance of rays from the centre to the circumference.

Umbri'nus. The colour of umber.

Umbro'sus (shady). Synonyme for Umbraticolus.

Unangula'tus (*unus* one, *angulus* an angle). When a stem, &c., has a projecting line or angle along one side only.

UNARMED. Without any sharp point at the apex. Without spines, prickles, or other sharp projections.

Unca'tus (*uncus* a hook). Hooked.

UNCERTAIN. Without determinate direction.

Un'cia (an inch). About an inch long.

Uncia'lis. Of the length expressed by Uncia.

Uncifor'mis, *Uncina'tus* (*uncus* a hook, *forma* shape) Synonymes for Uncatus.

UNCOVERED Synonyme for Naked.

Unctuo'sus (*unctus* greasy). See Greasy.

Un'cus. A Hook.

Unda'tus (formed with a wave-like aspect). Waved.

UNDER-SHRUB. A plant only partially shrubby, the ends of the newly-formed branches continuing herbaceous, and dying away in winter.

Undo'sus, Undula'tus (waved). Synonyme for Repandus. Synonyme for Undatus.

Unequal, Unequal-sided. When opposite sides are not symmetrical. Synonyme for "Irregular."

Unequally-pinnate. Pinnate, with an odd leaflet at the extremity; fig. 183.

Unguic'ulate, *Unguicula'ris Unguicula'tus* (*unguis* a nail). Furnished with a claw.

Un'guis (a nail). A "claw." Also, about the length of the finger nail, or half an inch.

Unicapsula'ris (*unus* one, *capsula* a capsule). A fruit composed of a single capsule.

Unicellula'ris (*unus* one, *cellula* a cell), Composed of a single cell.

Uni'color (*unus* one, *color* colour). Of one uniform tint.

Unicotyledo'neus (*unus* one, *κοτυληδων* a cotyledon). Has been employed synonymously with Monocotyledonus.

Uni'cus (one alone). Where there is only one of a particular part specified.

Uniembryona'tus (*unus* one, *embryo* the embryo). Where a seed contains (as most usually) only one embryo.

Uniflo'rus (*unus* one, *flos* a flower). Supporting, or subtending, a single flower.

Unifolia'tus, Unifo'lius (*unus* one, *folium* a leaf). Bearing only a single leaf.

Unifoliola'tus (*unus* one, *foliolum* a leaflet). Where a peduncle supports a single leaflet, distinguished as such by being articulate to it.

Unifora'tus (*unus* one, *foramen* a hole). Opening by a single hole or pore.

Unifor'mis (*unus* one, *forma* shape). When the receptacle in Compositæ bears florets of one description only.

Unigem'mius (*unus* one, *gemma* a bud). Giving origin to a single bud.

Unige'nus (*unus* one, *gigno* to produce). Putting forth leaves once only in the year.

Unijuga'tus, Uniju'gus (*unus* one, *jugum* a yoke). A pinnate form with only a single pair of subordinate parts.

Unilabia'tus (*unus* one, *labium* a lip). An irregular monopetalous corolla with only one lip. A monopetalous corolla slit on one side; as in the "ligulate" florets of Compositæ.

UNILAT'ERAL, *UNILATERA'LIS* (*UNUS* one, *LATUS* a side). Either, disposed along one side; or, entirely forming one side.

UNILOBA'TUS (*UNUS* one, *LOBUS* a lobe). Having a single lobe. Synonyme for Monocotyledonus.

UNILO'CULAR, *UNILOCULA'RIS* (*UNUS* one, *LOCULUS* a cell). With one cell only.

UNINERVA'TUS UNINER'VIS UNINER'VUS (*UNUS* one, *NERVUS* a nerve). Where there is only one nerve; or where only one is very distinctly perceptable.

UNINTERRUPTED. Synonyme for "Continuous."

UNIOCULA'TUS (*UNUS* one, *OCULUS* an eye). With only one vegetating point.

UNIOVULA'TUS (*UNUS* one, *OVULUM* an ovule). When a cell in the pericarp contains only one ovule.

UNIPA'ORUS. With one peduncle only.

UNIPET'ALUS (*UNUS* one, *πεταλον* a petal). Where only a single petal is produced, and does not surround the inner floral whorls; being used in contradistinction to gamopetalous and monopetalous.

UNISERIA'LIS UNISERIA'TUS (*UNUS* one, *SERIES* a row). Disposed in a single row, or in one whorl.

UNISEX'UAL, *UNISEXUA'LIS*, *UNISEX'US* (*UNUS* one, *SEXUS* a sex). A flower which has either stamens alone, or ovaries alone. A plant which bears only unisexual flowers.

UNIVAL'VIS (*UNUS* one, *VALVA* a valve). When a capsular fruit dehisces along a single suture.

UNIVERSAL, *UNIVERSA'LIS.* Synonyme for "General."

UNIVESICULA'RIS (*UNUS* one, *VESICULA* a vesicle). Synonyme for Unicellularis,

URCE'OLATE, *URCEOLA'RIS*, *URCEOLA'TUS* (*URCEOLUS* a little pitcher). Shaped somewhat like a pitcher with a contracted mouth; fig. 184.

URCE'OLUS (a little pitcher). A membranous or cartilaginous tube, swollen below, and more or less contracted above.

U'RENS (burning). Stinging. See Sting.

URN, *UR'NA.* The theca or spore-case of Mosses. The base of a Pyxidium.

URTICA'CEÆ, URTI'CEÆ (from the genus Urtica). The Nettle tribe. A natural order of Dicotyledones.

U'TERUS (the womb). Synonyme for Volva.

URTRI'CLE, *URTRI'CULUS* (a little bottle). A small, superior, membranous, and monospermous pericarp. Also, a little bladder filled with air, attached to certain aquatic plants. Synonyme for Urceolus, and for Vesicula.

UTRI'CULAR, *URTICULA'RIS*, *URTICULA'TUS* (*UTRICULUS* a little bottle). Synonyme for Inflatus.

UTRICULIFOR'MIS (*UTRICULUS* a little bottle, *FORMA* shape). Synonyme for Urceolatus.

UTRICULARI'NEÆ (from the genus Utricularia). Synonyme for Lentibulaceæ.

UTRICULO'SUS (*UTRICULUS* a little bottle). Bearing many of the air-bladders termed utriculi. Synonyme for Utricularis.

UTRIFOR'MIS, *UTRI'GERUS* (*UTRICULUS*, *FORMA* shape, and *GERO* to bear). Synonymes for Utriculiformis.

UVA'RIUS, *UVI'FERUS*, *UVIFOR'MIS* (*UVA* a grape, *FERO* to bear, and *FORMA* shape). Composed of round parts connected by a support, like a bunch of grapes.

VACCINA'CEÆ, VACCI'NIEÆ (from the genus Vaccinium). The Bilberry tribe. A natural order of Dicotyledones.

VACCI'NUS (belonging to a cow). Of a Dun colour.

VACIL'LANS (waving). Synonyme for Versatilis.

VA'CUUS (void). When an organ is without some part which is usually present within it, or with it: as a carpel without ovules, a bract without a flower bud.

VAGIFOR'MIS (*VAGUS* inconstant, *FORMA* shape). Possessing no well defined form.

VAGI'NA. A "sheath." Also any part which completely surrounds another.

VAGI'NANS. Assuming the condition of a Vagina.

VAGINA'TUS. Surrounded by a Vagina.

VAGINEL'LA (diminutive of Vagina). Where a Vagina is very small.

VAGINER'VIS, *VAGINER'VIUS* (*VAGUS* inconstant, *NERVUS* a nerve). Where the nerves are irregularly disposed, in various directions, as in the leaves of succulent plants.

VAGINIF'ERUS (*VAGINA* a sheath, *FERO* to bear). Furnished with one or more sheaths.

VAGIN'ULA (a little sheath). A small sheath at the base of the seta in Mosses. Synonyme for a tubular floret in Compositæ.

VA'GUS (wandering or inconstant). Proceeding in no definite direction.

VAIL. See Veil.

VALERIANA'CEÆ, VALERIA'NEÆ (from the genus Valeriana). The Valerian tribe. A natural order of Dicotyledones.

VALLE'CULA (diminutive from *VALLIS* a valley). A depressed space (interstice) between the primary "Ridges" on the fruit of Umbelliferæ.

VALVE, *VAL'VA* (*VALVÆ* doors). Distinct portions of certain organs (as in anthers and pericarps) which become detached by regular dehiscence along definite lines of suture.

VALVA'CEUS (*VALVÆ* doors). Furnished with valves.

VALVATE, *VALVA'RIS*, *VALVA'TUS* (with folding doors). When contiguous organs, or similar subordinate parts, touch each other along the edges without over-lapping; fig. 185.

VALVEA'NUS (*VALVA* valve). When a partition emanates from the expansion of the inner substance of a valve.

VAL'VULA (diminutive of *VALVA*). Used as a synonyme for Perithecium in some cases. Also for any of the floral bracts in Gramineæ. Any small valve-like expansion.

VAL'VULAR. Synonyme for Valvate.

VALVULA'TUS (from *VALVULA*). Synonyme for Articulatus in its application to cellular tissue.

VANILLA'CEÆ (from the genus Vanilla). A natural order of Monocotyledones.

VARIA'BILIS, *VA'RIANS* (varying). Presenting a variety in character; as when leaves are variously modified on the same plant, &c.

VA'RIEGATED, *VARIEGA'TUS*, *VA'RIUS* (changeable. Where colours are disposed in irregular patches.

VARI'ETY, *VARI'ETAS*. An individual possessing a form, to a certain degree modified from that which is considered to be most characteristic of the species.

VARIIFO'LIUS (*VARIUS* various, *FOLIUM* a leaf). Possessing leaves of different forms.

VARI'OLA (the pustule of small pox). A shield in the genus Variolaria, having a pustular appearance.

VA'RIUS (changeable). Where colour gradually changes from one tint to another.

VAS. A vessel.

VAS'CULAR, *VASCULA'RIS*, *VASCULO'SUS* (*VAS* a vessel). Containing vessels.

VAS'CULAR SYSTEM. Those interior portions of any plant in which vessels occur.

VAS'CULUM (a little vessel). Synonyme for Ascidium.

VASE-SHAPED. Shaped somewhat like a common flower pot without its rim.

VASIDUC'TUS (*VAS* a vessel, *DUCO* to lead). Synonyme for Raphe.

VA'SIFORM TISSUE. Synonyme for Ducts.

VAULTED. See Fornicatus.

VEIL. A membrane which invests the theca in Mosses; and which, by the growth of the seta and expansion of the theca, is ruptured and carried up upon the lid; fig. 186, *v* veil, *s* theca, *t* seta. Also, a membrane which invests the pileus, and is connected with the stipes in certain Fungi.

VEILED. Partly hidden.

VEIN. A bundle of fibro-vascular tissue penetrating a leaf or foliaceous appendage.

VEIN'LESS. Possessing no veins.

VEIN'LET. The smallest ramifications of a vein.

VELAMINA'RIS (*VELAMEN* a veil). When an anther dehisces by the rolling up of one side of a cell from base to apex.

VELA'TUS. Veiled.

VEL'LUS (a fleece). The stipes of some Fungi.

VE'LUM. The veil in certain Fungi.

VELU'MEN (from *VELLUS* a fleece). Velvet. A coating of close soft hair.

VELU'TINUS, *VELUTINO'SUS* (from *VELUMEN*). Velvety. With a surface resembling velvet, being coated with velumen.

VE'NA. A vein.

VENA'TION, *VENA'TIO*. The arrangement of veins.

VENENI'FERUS (*VENENUM* poison, *FERO* to bear). Producing poisonous matter.

VENO'SUS (*VENA* a vein). With numerous veins.

VENTILATO'RIUS (*VENTILABRUM* a fan). Synonyme for Flabellatus.

VEN'TRAL, *VENTRA'LIS* (*VENTER* the belly). Used in contradistinction to Dorsal. Thus, in a pericarp formed from a single carpel, the "ventral suture" would be the line of union between the placentiferous edges.

VENTRICO'SE, *VENTRICO'SUS* (big-bellied). Swelling out on one side.

VENTRICULO'SUS. Slightly ventricose.

VEN'ULA (a small vein). A veinlet.

VENU'LÆ-COMMU'NES. Anastomosing veinlets.

VENULO'SO-HINOI'DEUS (*VENA* a vein, and ινοειδης with nerves). When equal and curved veins proceed parallel to each other from the midrib to the margin.

VENULO'SO-NERVO'SUS (*VENA* a vein, and *NERVUS* a nerve). When straight parallel veins are connected by cross veinlets.

VERBENA'CEÆ (from the genus Verbena). The Vervain tribe. A natural order of Dicotyledones.

VER'DIGRIS-GREEN. Deep bluish green. Yellow and blue, the latter in excess.

VERMICULA'RIS (*VERMICULARE* worm-like). Worm-shaped.

VERMICULA'TUS (infested with worms). Covered with contorted worm-like elevations. Synonyme for Miniatus.

VER'NAL, *VERNA'LIS*, *VER'NUS* (belonging to Spring). Appearing at spring time.

VERNA'TION, *VERNA'TIO* (a renewing). The manner in which leaves are disposed in the bud.

VERNICO'SUS (*VERNIX* varnish). When a surface appears polished, as if by varnish.

VERRU'CA. A wart. Also the perithæcium of some Fungi.

VERRU'CÆFORM, *VERRUCÆFOR'MIS* (*VERRUCA* a wart, *FORMA* shape). Resembling a wart.

VERRUCO'SUS. Warty.

VERRUCULO'SUS (*VERRUCULA* a little wart). Where the warts are small and abundant.

VER'SATILE, *VERSAT'ILIS.* When a part is so slightly attached to its support that it readily swings to and fro.

VERSIC'OLOR (changing colour). Possessing several tints of colour. Or, appearing differently coloured in different positions.

VERSIFOR'MIS (*VERSO* to turn, *FORMA* shape). Changing its shape as it grows old.

VERSIPAL'MUS (*VERSO* to turn, *PALMA* the hand). A palmate arrangement, in which the divisions are not all in the same plane.

VER'TEBRATE, *VERTEBRA'TUS* (in the form of a vertebra). Distinctly articulated, and often more or less contracted at intervals.

VER'TEX (the top). Any upper extremity. The pileus of certain Fungi.

VER'TICAL, *VERTICA'LIS.* When the axis of any part is perpendicular to any other from which it arises. Also, in the usual sense of a direction perpendicular to the horizon.

VER'TICEL, *VERTICIL'LUS* (*VERTICILLUM* a little whorl, from *VERTO* to turn). A whorl.

VERTICIL'LASTER (*VERTICILLUS* a whorl, *ASTER* a starwort). When short cymes in the axils of opposite leaves give to the inflorescence of Labiatæ the appearance of their flowers being disposed in whorls; fig. 187.

VERTI'CILLATE, *VERTICILLA'TUS* (*VERTICILLUS* a whorl). Whorled.

VERTICILLIFLO'RUS (*VERTICILLUS* a whorl, *FLOS* a flower). When whorls of flowers have a spiked arrangement.

VERTICILLA'TO-PINNATISEC'TUS (*VERTICILLATUS* whorled, *PINNATISECTUS* pinnately divided). When certain sessile leaves are subdivided into numerous filiform pinnately-arranged segments, which assume an appearance as if they were whorled about the stem.

VERTICIL'LUS. See Verticel.

VERTICIL'LUS-SPU'RIUS (*SPURIUS* counterfeit). False whorl. Synonyme for Verticillaster.

VERUCULA'TUS (*VERUCULUM* a little broach). Cylindrical and somewhat pointed.

VESIC'ATORIUS (*VESICA* a bladder). Producing blisters when applied to the skin.

VE'SICLE, (*VESICULA* a little bladder). A bladder-like cavity filled with air.

VESI'CULAR, *VESICULÆFOR'MIS*, *VESICULA'RIS*, *VESICULA'TUS*, *VESICULO'SUS* (*VESICULA* a little bladder, *FORMA* shape). Bladdery.

VESICULIF'ERUS (*VESICULA* a little bladder, *FERO* to bear). Supporting or containing bladders.

VESPERTI'NUS (of the evening). Appearing or expanding in the evening.

VES'SEL. A cell which assumes a lengthened tubular condition. See Duct and Tracheæ.

VEXIL'LARY, *VEXILLA'RIS* (*VEXILLUM* a standard). The arrangement of the petals in the æstivation of a papilionaceous flower; fig. 188.

188

VEXILLA'TUS (*VEXILLUM* a standard). When a papilionaceous flower has a large standard.

Vexil'lum. A standard. See Papilionaceous.

Vice'ni (twenty). In twenties together.

Vigi'liæ (*vigilia* a watching). Applied to the periods during which certain plants gradually expand and close their flowers daily.

Villo'se, *Villif'erus*, *Villo'sus* (*villus* wool, *fero* to bear). Covered with long weak hair.

Villos'ity, (*Villus* wool). A covering of long weak hair.

Vi'men (a twig). A long flexible shoot.

Vimi'neous, *Vimi'neus* (made of wickers). Furnished with long flexible twigs.

Vine. Any trailing stem resembling that of the Grape-vine.

Vinea'lis (belonging to a vine and vineyard). Growing naturally in vineyards.

Vini'feræ (*vinum* wine, *fero* to bear). Synonyme for Vitaceæ,

Vino'sus (having the savour of wine). Of the colour of red wine. Dirty pale red. Red with much grey.

Viola'ceæ, Viola'rieæ (from the genus Viola). The Violet tribe. A natural order of Dicotyledones.

Violaces'cens. With a pale tinge of violet.

Vi'olet, *Viola'ceus.* Of a violet colour. Blue with a little red.

Vi'rens. Green.

Vires'cens. Somewhat green.

Vir'gate, *Virga'tus.* Twig-like.

Virgin'eus (virgin-like). Having attained the state of flowering.

Virgul'tum. A twig.

Virides'cens (*viridis* green). Synonyme for Virescens.

Viridi'na (*viridis* green). Synonyme for Chlorophylla.

Vi'ridis. Green. Also Viridulus.

Vi'ror. Greenness.

Viro'sus (venemous). With a noisome smell.

Vis'cid, *Vis'cidus*, *Visco'sus.* Coated with a tenacious juice.

Viscoi'deæ (*viscum* bird-lime). Synonyme for Loranthaceæ.

Vita'ceæ, Vi'tes (from the genus Vitis). The Vine tribe. A natural order of Dicotyledones.

Vitel'linus (*vitellus* the yolk of an egg). The colour of the yolk of an egg. Orange with a little grey.

Vitel'lus (the yolk of an egg). The thickened sac within the nucleus which contains the amnios. Has been also described as any portion attached to the embryo, not distinctly referable to radicle, cotyledon, or plumule. An oily substance adhering to the spores of Lycopodiaceæ.

Viti'colus (*vitis* a vine, *colo* to inhabit). Living on or within the Vine.

Viti'cula (a little vine). Synonyme for Surculus.

Viticulo'sus. Producing viticulæ.

Vi'treus. Transparent.

Vi'tricus (*vitrum* glass). Looking like glass.

Vit'ta (a fillet). A narrow elongated receptacle of aromatic oil, of which there are often several longitudinally and regularly disposed in the spermoderm, in Umbelliferæ; fig. 189. In a transverse section of the fruit they appear as brown dots between the pericarp and albumen.

Vitta'tus (*vitta* a fillet). Striped longitudinally.

Vochya'ceæ, Vochysia'ceæ (from the genus Vochya or Vochysia). A natural order of Dicotyledones.

Volu'bilis (easy to be turned). Twisting spirally round a support.

Volu'tus (rolled). Rolled up in any direction.

Vol'va (a wrapper). A membrane which completely invests certain Fungi in their early stages, and which bursts open as the contents develop.

Wart. A firm glandular excrescence, or hardened protuberance on the surface.

Warty. Covered with warts.

Waved, Wavy. Having an alternately convex and concave surface or margin.

Waxy. Resembling bee's wax in texture and colour. Yellow, a little dulled by grey.

Wedge-shaped. Approaching an isosceles triangle with narrow base, the point of attachment being at the apex; fig. 190.

Wheel-shaped. See Rotate.

Whip-shaped. See Flagelliform.

White. Without any tinge of colour.

Whitened. When a slight covering of white shows a darker ground beneath.

Whi'tish. Not pure white.

WHORL. Any set of organs or appendages arranged in a circle round an axis, and in, or very nearly in, a plane perpendicular to it.

WHORLED. Disposed in whorls.

WILD, Growing without cultivation, whether indigenous or naturalized.

WING. A membranous expansion. Each of the two lateral petals in a papilionaceous or some other irregular flower in which these differ from the rest.

WINGED. Having wings.

WINTE'REÆ (from the genus Wintera or Drimys). The Winters-Bark tribe. A natural order of Dicotyledones.

WITHERING. Synonyme for Marcescent.

WOOD. The inner and hardened portions of stems of more than one year's duration.

WOODY. Becoming or approaching the nature of wood.

WOOLLY. When hairs are long, curled, and matted together like wool.

WORM-SHAPED. More or less cylindrical and contorted.

WRINKLE. An irregular elevation of one surface with a corresponding indentation of another, on opposite sides of a lamina.

WRINKLED. Disposed in wrinkles.

XANTHOPHY'LL, *XANTHOPHYL'LUM* (ξανθος yellow, φυλλον a leaf). A yellow colouring matter in plants.

XANTHOXYLA'CEÆ, XANTHOXY'LEÆ (from the genus Xanthoxylum). A natural order of Dicotyledones.

XERAMPELI'NUS (somewhat ruddy). A very dull brown red. Red with much grey.

XI'PHOPHYLLUS (ξιφιον a sword, φυλλον a leaf). Having ensiform leaves.

XY'LINUS (ξυλον wood). Used as a synonyme for Lympha.

XYLOCAR'PUS (ξυλον wood, καρπος fruit). When fruit becomes hard and woody.

XYLO'DIA, *XYLO'DIUM* (ξυλον wood, ειδος like). The fruit of Anacardium; unsymmetrical, monospermous, woody, and seated on a fleshy support. Also synonyme for Achenium.

XYLOMY'CES (ξυλον wood, μυκης fungus). Fungi which grow on wood or bark.

XYRIDA'CEÆ, XYRI'DEÆ (from the genus Xyris). A natural order of Monocotyledones.

YEARLY. Of a year's growth.

ZANTHOXY'LEÆ. Synonyme for Xanthoxylaceæ.

ZINGIBERACEÆ, ZINZIBERA'CEÆ (from the genus Zingiber). Synonyme for Scitamineæ.

ZOO'CARP, *ZOADU'LA*, ZOOSPER'MA (ζωον an animal, αδηλος doubtful, σπερμα seed). The spores of certain Algæ, which are for a time endowed with powers of locomotion.

ZYGOPHYLLA'CEÆ, ZYGOPHYL'LEÆ (from the genus Zygophyllum). The Bean-Caper tribe. A natural order of Dicotyledones.

Printed in the United States
By Bookmasters